WILDLAND FIRE APPARATUS 1940 - 2001
PHOTO GALLERY

John H. Rieth

Iconografix
Photo Gallery Series

Iconografix
PO Box 446
Hudson, Wisconsin 54016 USA

© 2001 John H. Rieth

All rights reserved. No part of this work may be reproduced or used in any form by any means... graphic, electronic, or mechanical, including photocopying, recording, taping, or any other information storage and retrieval system... without written permission of the publisher.

The information in this book is true and complete to the best of our knowledge. All recommendations are made without any guarantee on the part of the author or Publisher, who also disclaim any liability incurred in connection with the use of this data or specific details.

We acknowledge that certain words, such as model names and designations, mentioned herein are the property of the trademark holder. We use them for purposes of identification only. This is not an official publication.

Iconografix books are offered at a discount when sold in quantity for promotional use. Businesses or organizations seeking details should write to the Marketing Department, Iconografix, at the above address.

Library of Congress Card Number: 2001131942

ISBN 1-58388-056-9

01 02 03 04 05 06 07 5 4 3 2 1

Printed in the United States of America

Cover and book design by Shawn Glidden

Copyedited by Dylan Frautschi

BOOK PROPOSALS

Iconografix is a publishing company specializing in books for transportation enthusiasts. We publish in a number of different areas, including Automobiles, Auto Racing, Buses, Construction Equipment, Emergency Equipment, Farming Equipment, Railroads & Trucks. The Iconografix imprint is constantly growing and expanding into new subject areas.

Authors, editors, and knowledgeable enthusiasts in the field of transportation history are invited to contact the Editorial Department at Iconografix, Inc., PO Box 446, Hudson, WI 54016.

DEDICATION

This book is dedicated to my father Henry E. Rieth, ex-Chief of the Atlantic Highlands, New Jersey, Fire Department. His dedication and love for the fire service was inherited by myself! Because of his encouragement and support, I have a deep appreciation for fire apparatus as well as a career in the fire service. Thank you Dad!

ACKNOWLEDGMENTS

My appreciation to all the fire apparatus photographers who sent me over 2,000 photographs to pick from for this book. Thanks go out to several manufacturers—John Giesfeldt of Pierce Manufacturing, Lee Egart of Lee's Emergency Equipment, Gary Burton of G&S, and also to the S&S Fire Apparatus Company.

This book could not have been possible without the work of these gifted and talented fire apparatus photographers. The list seems to be a who's who list of fire apparatus photographers, as follows: Dennis J. Maag, Dick Bartlett, Hank Knight, Rick Davis, Jack Wright, Ron Jeffers, Mark V. Carr, Adam Alberti, Joe Abrams, Daniel A. Jasina, David Bowen, Garry Kadzielawski, Robert P. Vaccaro, Chuck Madderom, Howard Meile, Patrick Shoop, Robert W. Fitz Jr., Mike Tenerelli, Glenn B. Vincent, Mark A. Redman, Gregory Stapleton, Shaun P. Ryan, Stefan Farage, John Rowe, Deran Watt, Dennis C. Sharpe, Mike Sanders, Todd Lincoln, Mike Defina Jr., John Malecky, John M. Calderone, Frank Wegloski, Bob Allen, Alan Mudry, and John D. Floyd.

These next folks not only lent me photographs but helped with picking photos, and doing research as well: Tim Regan, Scott Mattson, John Rowe, Bill Schwartz, and the late Jim Burner.

Some others that I would like to thank are Mike Martinelli, Tom Finnegan, Jim Hill of North Tree Fire Service, Dave Schlosser, John A. Calderone, as well as Maris Gabliks, David Harrison, and Chris Irick—all from the New Jersey Forest Fire Service.

A special thank you to my beloved wife Michele, for all the hours I spent away from her when I was working on this book, and to my mother Marion and grandmother Juile.

Please note several people sent me the same photo of the same truck. I picked the photos based on several factors. I tried to be as fair as possible. Thank you to everyone for supplying me with over 2,000 photographs to choose from.

INTRODUCTION

Wildland Firefighting Apparatus have been around since motorized vehicles were developed. Wildfire fighting equipment is used everywhere from large, urban cities to the smallest of rural communities. With a variety of names, functions, and manufacturers/custom builders, these apparatus provide a fire buff/fire apparatus historian with an interesting and different perspective from one rig to the next.

There are many different kinds of apparatus used to fight wildfires—units that attack the wildfire, water/tankers to supply water, transport trucks that haul heavy equipment to wildfires, and other support units like fuel trucks, repair trucks, crew transport units, and command and communication units. Other equipment like bulldozers, special tracked vehicles, and even aircraft are all used to help fight wildfires. Many fire departments and agencies have even had members convert old military or commercial trucks into wildland firefighting apparatus.

Throughout the years new apparatus have been developed to help fight wildfires. For instance, the wildland-urban interface engine was developed to provide a versatile vehicle equipped with more equipment and more ability than a smaller brush truck or a straight wildland engine. The interface engine's first job is to protect structures from wildfires. These rigs are more like the Midi pumpers of the 1980s, which bridges the gap between a 1-ton chassis and a full sized class A pumper. Some fire departments have set up their interface engines with more structural firefighting equipment than wildland fire equipment. Either way, these rigs should be able to handle either type of fires, but its main mission is structural protection.

Both large and small apparatus manufacturers have built apparatus used for wildland firefighting. Such manufacturers like Howe, Young, Hahn, American LaFrance, Seagrave, Pierce, S&S, as well as small regional or local builders. Many of the smaller manufacturers are seldom heard of except by people who live in the local area.

There are many different government agencies that either fight wildfires or help support them. The US Forest Service is the largest. The US Bureau of Land Management, the US National Park Service, the US Fish and Wildlife Service and the Bureau of Indian Affairs are others. Many large government and military installations have wildland fire apparatus as well. Every state has some type of agency that either operates wildfire suppression apparatus, or assigns apparatus to local fire departments. Some states just provide heavy equipment or fire managers called overhead personnel. In addition, many county and municipal fire departments also operate wildland fire apparatus.

Different regions of the nation use various styles of wildland fire apparatus. Even different parts of the same state have different styles of apparatus for fighting wildfires. The terrain, elevation, and the type of vegetation are all considered when building a wildland fire apparatus. Some areas call their wildland fire apparatus brush trucks, field trucks, grass trucks, mud buggies, brush-breakers, stump-jumpers, wildland engines, or even brush engines.

Wildland fire apparatus are some of the more diverse types of fire apparatus to be seen. Through this chronological pictorial overview, I hope you discover how interesting these apparatus are.

The California Department of Forestry is one of the largest fire suppression agencies in the US. Not only is the CDF mission wildland firefighting, but municipal/county structural fire suppression is performed under contract for many communities and counties throughout the state. In Nevada City, California, the CDF keeps a 1929-1931 Ford (AAA) California Highway wildland engine as a muster/show rig. This rig has a 65-gpm pump and a 300-gallon tank. *Photo by Garry Kadzielawski.*

This circa 1939 Ford was placed into service by the New Jersey Forest Fire Service. At the time the New Jersey Forest Fire Service was in the Department of Conservation and Development. An unknown builder built the body. Note the buckets and backpack pumps used to fight wildfires. *Photo by Maris Gabliks, NJ Forest Fire Service.*

Some old trucks never die. This circa 1940 Chevy chassis had a locally built body placed on it. Several of these bodies were built for the New Jersey Forest Fire Service. Some of these bodies were mounted on the Chevy chassis and others were placed on other chassis. Believe it or not, several of these bodies are still in service with the New Jersey Forest Fire Service (some are mounted on their third chassis). *Photo by Maris Gabliks, NJ Forest Fire Service.*

The New Jersey Forest Fire Service operated this 1940-1941 Ford pickup that was fitted with a small water tank and hose reel in the cargo bed. A small front-mounted pump was installed as well. *Photo by Maris Gabliks, NJ Forest Fire Service.*

This circa 1941-1942 Dodge 1/2-ton 4x4 WC model paved the way for the famous Dodge Power Wagons that were produced for the civilian market after World War II. This Dodge WC is believed to be a WC4 and was acquired by the New Jersey Forest Fire Service to tow this circa 1946 Oliver bulldozer/plow unit. *Photo by Maris Gabliks, NJ Forest Fire Service.*

Forked River Fire Company of Lacey Township, New Jersey, at one time operated two of these International Model M-3H-4s. It was originally built for the Navy/Marines during World War II. This particular unit was a crash truck, of which 210 were built from 1941-1944. The 4x4 unit had a small pump and small tank (approximately 200 gallons). *Photo by Bill Schwartz, Jim Burner Collection.*

During World War II International Harvester built thousands of the M-5-H6 2 1/2-ton 6x6s for the Marine Corps between 1941-1945. This M-5-H6 was converted into a forestry unit for the Newfield, Maine, fire department. This truck named "Bertha" has a 250-gpm pump and a 500-gallon water tank. *Photo by Frank Wegloski.*

The Louden Fire Company of Waterford Township, New Jersey, retrofitted this circa 1943 Dodge WC(Weapons Carrier)63 weapons carrier with a cab that came off of a much newer civilian Dodge Power Wagon. The WC62 and WC63 used 1 1/2-ton tandem axle chassis, with over 23,000 produced between 1942-1945. This unit carried a 200-gpm pump and a 300-gallon water tank. The company named this truck "Wild Cherry." *Photo by Scott Mattson.*

In Bayville, New Jersey, the Pinewald Pioneer Forest Fire Fighters operated this World War II Chevy 4x4. One note about military models is that each model could have various variants, which means that different vocations or trucks made for a particular job (like a wrecker) would have a different model number, but the truck was based on the original model. This circa 1942 Chevy was based on the G7107 chassis, which had more than seven variants. In addition, two tactical fire apparatus were constructed on this G7107 chassis. The class 300 was intended for brush fires and the class 325 was set up for structural fires. They both were very similar. Both classes came equipped with a 300-gpm front-mounted pump and a 300-gallon water tank. The bodies were manufactured by Darley, Maxim, and General Detroit for the class 325, and the class 300 had fire equipment manufactured by Central Fire Truck Co. *Photo by Bill Schwartz, Jim Burner Collection.*

Farmingdale, New Jersey, Volunteer Fire Company received this circa 1942 Chevy CCKW 2 1/2-ton chassis. After World War II the War Assets Administration distributed surplus vehicles to state, county, and local agencies. Some of the trucks that fire department/companies received were tactical units, which are used in combat situations, and generally look like military trucks. Some fire apparatus were manufactured for the military on these tactical chassis. When local fire companies started to receive these units they would accept any type. Some were tactical cargo trucks that had to be converted into fire apparatus. Across America, many fire companies' and departments' first motorized apparatus came from Government surplus. This unit was converted by the volunteer members of the fire company; they had equipped it with a 500-gpm pump and a 1000-gallon tank. *Photo by Bill Schwartz, Jim Burner Collection.*

The Jamesburg Fire Company converted this Dodge WC51 into a brush truck. It had a 200-gallon tank and a small 100-gpm pump, and was in service until the late 1980s. The Dodge WC51 3/4-ton chassis was manufactured for the military from 1942-1945. *Photo by Tim Regan.*

The Rivera Beach Volunteer Fire Company of Anne Arundel County, Maryland, ran this ex World War II half-track as a wildland firefighting unit. The company was nicknamed the "Son's of the Beach." This truck was modified with a closed cab and reconfigured for firefighting. The pump size is not known, nor is the tank. The half-tracks were built by White, Diamond T, and Autocar from 1940-1944. From 1944-1945 International produced the half-track. Other fire departments used two-ton half-tracks, including Philadelphia, Pennsylvania, Fire Department. *Photo by Howard Meile.*

The EastPort Volunteer Fire Department of Annapolis, Maryland, ran this 1943 International M-3L-4 ex military crash truck. It has a 250-gpm pump and a 200-gallon tank. *Photo by Howard Meile.*

Quaker Springs, New York, operates this 1944 International M-3L-4 ex military crash tender. This unit has a Bean high-pressure pump and a 400-gallon water tank. *Photo by Frank Wegloski.*

During World War II, the military had several tactical fire apparatus built. This is a 1945 Chevy class 525 pumper. It was based on the Chevy G7107 1 1/2-ton 4x4 chassis. As delivered to the Army, it had a 500-gpm front-mounted pump and a 150-gallon booster tank. The bodies were built by Oren. Today, the Forksville, Pennsylvania, Volunteer Fire Department operates this unit fitted with a 250-gallon tank. *Photo by Todd Lincoln.*

Besides the CDF, the US Forest Service also operates one of the largest fire suppression forces in the nation. This 1947 Ford with a 4x4 system by Marmon-Herrington has been restored by a firefighter with the US Forest Service. The truck is used for fire prevention, musters and parades throughout the Cleveland National Forest area. The truck is out of the El Cariso Station in California. *Photo by Chuck Madderom.*

Frank Wegloski provided this photo of Bedford, Massachusetts's 1948 Ford Farrar 250-gpm forestry unit. Notice the lack of four-wheel drive, but the front-mounted winch is included.

The Sudlersville, Maryland, Fire Company had this circa 1950 Jeep 4x4 pickup in service. This unit had a small pump and tank. Jeep produced this type of truck from 1947-1965 and very few notable changes were made. *Photo by John D. Floyd.*

This circa 1950 Oliver dozer has an unknown trailer-mount device. The dozer was a New Jersey Forest Fire Service unit. *Photo by NJ Forest Fire Service.*

This GMC military M-211 6x6 was refurbished during the 1990s. The Almoson Lake Fire Company of Deptford Township, New Jersey, not only uses this attractive unit for brush fires, but also has it equipped with an A-frame for rescue operations. Note the stokes basket on top of the 1000-gallon water tank, which can be used for removing the injured. The M-211 2 1/2-ton chassis with its trademarked, short, slanted hood was produced only by GMC in seven different variants. They were manufactured from 1950-1955. *Photo by Scott Mattson.*

This M-54 model 5-ton tactical chassis came in 21 different variants, each with a different model number. The Centerton, New Jersey, Fire Company converted this military Diamond T into a fire apparatus. Many rural fire companies could not afford both wildland and structural fire apparatus, so they would build a truck that could somehow function a little as both. This unit was believed to carry a 500-gpm pump, and held 500 gallons of water. The M-54 models were manufactured by International, Diamond T, Mack, and Kaiser (AM General) from the 1950s to the late 1970s. *Photo by Bill Schwartz, Jim Burner Collection.*

This 1951 Studebaker is in service with the Wiley Ford, West Virginia, Fire Department. The body was manufactured by an unknown builder. The bumper is lettered as Forest Fire Unit. *Photo by Howard Meile.*

This attractive red and white M-211 GMC 2 1/2-ton 6x6 is in service as Forestry Tanker 10 of the Westfield, Connecticut, Fire Department. The M-211 chassis had seven variants, and all were produced by GMC from 1950-1955. These trucks were the only trucks that ever served in the military to have this short, sloped hood. *Photo by Glenn B. Vincent.*

On Cape Cod the standard wildfire fighting truck is called a brush-breaker. On Long Island, New York, they are called stump-jumpers. After this type of truck pushes over a tree, it then has to "jump" over the stump of the tree. This 1952 GMC M-211 is in service with the Montauk, New York, Fire Department. It has a 500-gpm pump and a 500-gallon water tank. Note that units that had only single wheels on their rear axles were considered an M-135. Many Long Island fire departments turned dual wheels into single rear wheels for better traction off road. *Photo by Frank Wegloski.*

In Penn Forest Township, Pennsylvania, the Albrightsville Fire Company operates this 1952 GMC M-211 2 1/2-ton 6x6. This unit has a 500-gpm pump and a 500-gallon tank. *Photo by Todd Lincoln.*

Colts Neck, New Jersey, Fire Co. #1 received this 1952 Reo M-35 6x6 2 1/2-ton truck in the mid-1980s from the Colts Neck Fire Co. #2. The truck was completely refurbished and is still in service. This unit has a high-pressure pump and carries 500 gallons of water. *Photo by John Rieth.*

The Hooksett, New Hampshire, Fire Department has in service this 1952 Dodge M-37 4x4, equipped with a 200-gpm pump and a 250-gallon tank. This truck is owned by the New Hampshire Forest Fire Service, but is operated by the Hooksett Fire Department. *Photo by Glenn B. Vincent.*

The Colchester, Connecticut, Fire Department runs this 1952 Dodge M-37 4x4, equipped with a 250-gpm pump and a 170-gallon tank. Note the military cargo bed has been replaced with a flatbed body, and brush protection has been added. *Photo by Glenn B. Vincent.*

West Annapolis Volunteer Fire Department's Brush 40 is a 1952 Reo M-35 2 1/2-ton. The modified body holds a 250-gpm pump and a 750-gallon tank. *Photo by Howard Meile.*

Ridge, Long Island, New York, ran this 1952 Ford/American 500-gpm 4x4 pumper equipped with a front-mounted pump and a 1000-gallon tank. *Photo by Frank Wegloski.*

Barnstable County Forest Fire Service, Massachusetts, at one time operated this circa 1953 Ford COE 6x6 brush-breaker. Not much information is known about this unit. We do know this chassis was produced from 1950-1956. *Photo by Frank Wegloski.*

Breton Woods Fire Co. of Brick, New Jersey, has converted this 1953 Reo M-35 2 1/2-ton 6x6. The Cambodian Cadillac (the company's nickname for this truck) is equipped with a 250-gpm pump and a 500-gallon tank. The screen behind the driver's door helps protect the firefighters who ride behind the cab when fighting fires; they use the short lengths of booster hose like the one laying on the step just ahead of the rear wheels. *Photo by John Rieth.*

The Chicopee Fire Department of Buxton, Maine, operates this 1953 Reo M-35 2 1/2-ton 6x6. This unit is equipped with brush bars, a 100-gpm pump, and an 825-gallon tank. *Photo by Frank Wegloski.*

Based on the Dodge Power Wagon, Dodge manufactured thousands of the M-37 3/4-ton 4x4s. Production started in 1950 and lasted into 1954, then production resumed in 1958 and continued until 1968. This 1953 M-37 has a high-pressure pump and a 150-gallon tank. It is in service with the Lebanon, Maine, Fire Department. *Photo by Frank Wegloski.*

This 1953 Chevy chassis had a locally built body placed on it. Tank 1 of the Goodwins Mills, Maine, Fire Department has a 200-gpm pump and a 600-gallon tank. This fire department provides service to Dayton and Lyman, Maine. *Photo by Frank Wegloski.*

Dodge manufactured thousands of the M-37 3/4-ton 4x4s, which replaced the older Dodge WC models in the military. Thousands of fire agencies have built firefighting apparatus on these M-37 chassis. This 1953 M-37 operated by the Oxford, New Jersey, Fire Department, is unusual due to the brush bars. Oxford, located in northwestern New Jersey, has more mountainous terrain than central and southern New Jersey, thus most northern New Jersey fire departments do not have such brush protection. This unit is equipped with a 150-gpm pump and a 20-gallon tank. The original M-37s were produced from 1950-1954, and then an updated M-37-B1 series was produced from 1958-1968. *Photo by John Malecky.*

Plainville, Indiana, Volunteer Fire Department runs this 1953 GMC M-211 with a strange paint job—it is red over yellow with a black hood. This unit has a 100-gpm pump and a 1000-gallon tank. *Photo by Frank Wegloski.*

After the large wildfires in 1997, hundreds of new stump-jumpers have been placed into service throughout Long Island. Yaphank Fire Department runs this 1953 M-211/M-135 with a 500-gpm pump and a 640-gallon tank. *Photo by Frank Wegloski.*

In 1980 the Kent Volunteer Fire Department of Kent Cliffs, New York, converted this 1953 Dodge M-37 4x4 into a wildland firefighting unit. This truck carries a 50-gpm pump and a 150-gallon water tank. *Photo by Frank Wegloski.*

The Colorado State Forest Service assigns wildland fire apparatus to county and municipal fire departments to fight wildfires throughout the state. Here we see a 1954 GMC M-211 2 1/2-ton 6x6 operated by the Basalt & Rural Fire Protection District in Basalt, Colorado, for the Colorado State Forest Service. This engine has a 250-gpm pump, 1000-gallon water tank, and a 5-gallon class A foam tank. It serves the Pitkin and Eagle Counties area. *Photo by Dennis J. Maag.*

This 1954 FWD brush tanker was one of several built for the US Navy. These units had a 240-gpm pump and a 400-gallon tank. This photo was taken after its naval service. *Photo by John D. Floyd.*

This Maine Forest Fire Service rig was photographed by Frank Wegloski. Little information is known about this unit. It appears to be a circa 1954 International 4x4 chassis with a locally manufactured body.

At one time the East, and especially New Jersey fire agencies, operated hundreds of the classic Dodge Power Wagons. In fact, in southern New Jersey, the term Power Wagon is used to identify a brush truck. This circa 1955 Dodge WM 300 was operated by the Atco Fire Co. of Waterford Township, New Jersey. This unit was refurbished in the early 1990s. *Photo by Scott Mattson.*

This circa 1955-1957 International 4x4 was operated by the Marlboro, New Jersey, Fire Co. #1. The fire apparatus outfitter is unknown, however, this unit had a front-mount winch and racks for fire gear. *Photo by Bill Schwartz, Jim Burner Collection.*

Penns Creek, Pennsylvania, Fire Department operates this 1955 M-211. It has a 250-gpm front-mounted pump and a 1000-gallon tank. In addition, this rig has a portable, collapsible drop tank located between the tank and the cargo body on the driver's side (note the framing visible in photo). *Photo by David Bowen.*

Mashpee, Massachusetts, operated this 1957 International 4x4/Maynard brush-breaker. This unit carried a 750-gpm pump and a 750-gallon tank. Several local New England Fire Apparatus manufacturers have produced brush-breakers. *Photo by Robert W. Fitz Jr.*

The Tennessee Forestry Department assigned this 1957 International M-54 to the Caton's Chaple/Richardson Cove, Tennessee, Volunteer Fire Department. T-4 has a 500-gpm pump and a 2300-gallon tank. *Photo by Hank Knight.*

Trvro Township, Ohio, Division of Fire has this 1958 Chevy 4x4 pickup in service. Sutphen Fire Apparatus outfitted the truck with a 150-gpm pump and a 150-gallon tank. *Photo by Daniel A. Jasina.*

This well kept 1958 Chevy 4x4 has fire equipment by Towers Fire Apparatus. This truck formerly served the Waterloo, Illinois, fire department. The truck is now operated by the Maeystown, Illinois, Fire Department, who refurbished the truck in 1993. It has a 250-gpm pump and a 150-gallon tank. *Photo by Dennis J. Maag.*

This 1958 Reo M-35 was originally operated under a co-op agreement with the New Jersey Forest Fire Service and the Point Pleasant Boro Fire Co. #2. In the late 1980s, this 250-gpm pump and 500-gallon tank equipped truck was returned to the New Jersey Forest Fire Service and was reassigned to the Lonoka Harbor Fire Co. of Lacey Township, New Jersey, until its demise in the mid 1990s. *Photo by Scott Mattson.*

Dodge started to offer their Power Wagons with an optional conventional cab, as opposed to the WM military-looking cab, in 1957. This 1958 Dodge W-500 4x4 Power Wagon has a Seagrave body. This unit has a 250-gpm pump and a 500-gallon booster tank and was in service with the Kennebunk, Maine, Fire Department. *Photo by Frank Wegloski.*

The Robertsville Volunteer Fire Company of Marlboro Township, New Jersey, at one time operated two Dodge WM 300 Power Wagons. This 1958 Dodge is still in service. It is equipped with a 250-gpm pump and a 300-gallon tank. Company members had converted this unit from a pickup truck. Note the brush bars running along side and over the truck. During the 1960s the New Jersey Forest Fire Service began to outfit their brush trucks with this brush protection. This allowed the unit to cut a path through brush and forest to get to fires with minimal amount of damage to the truck. Many municipal fire companies in New Jersey used the New Jersey Forest Fire Service's brush bar styles on their own trucks. *Photo by John Rieth.*

This 1959 Diamond T M-54 5-ton 6x6 is operated by the Pinewald Pioneer Fire Company of Berkeley Township, New Jersey. This unit carries a 250-gpm pump and a 1000-gallon water tank. In New Jersey there are few fire departments/companies that operate 5-ton 6x6s. This model M-54 had twenty variants, and were manufactured from the early 1950s to the later part of the 1970s, by International, Diamond T, Mack, Kaiser/AM General. *Photo by John Rieth.*

Oceanic Fire Company of Staten Island, New York, is one of the few volunteer fire companies in New York City. This 1959 Dodge M-37 has a small pump and tank. Note the boat anchor to help winch the truck out of sand near the shoreline. *Photo by John M. Calderone.*

Mr. Lee Egart, who is a volunteer member of the West Tuckerton Fire Company of Little Egg Harbor, New Jersey, took this photograph shortly before retiring this circa 1960 M-35 2 1/2-ton 6x6. Mr. Egart would later start his own fire truck repair/refurbishing shop that would produce dozens of brush trucks, several that can be seen later in this book. The long-running M-35 series was produced from 1953 until the late 1970s. Manufacturers included Reo (the first was a Reo known as the Reo Eager Beaver), Studebaker, White, Kaiser-Jeep (later known as AM General). Thousands were built.

Mapaville, Missouri, uses this 1960 International 4x4. The members of the fire department mounted a 250-gpm pump and a 220-gallon tank in the pickup bed. *Photo by Dennis J. Maag.*

Tinton Falls Fire Co. #1 converted this 1962 Reo 6x6 M-35 into a brush fire unit, equipped with a flame painted on the hood, a 150-gpm pump and a 375-gallon tank that was installed on the truck. *Photo by John Rieth.*

The Potterville Fire Department of Scituate, Rhode Island, operated this open cab 1962 International 6x6 750-gpm pump equipped with an 800-gallon tank. The body was built by Maynard. *Photo by Glenn B. Vincent.*

In Jefferson County, West Virginia, the Wildfire Control Unit operates Forestry 19, a 1962 White M-35 2 1/2-ton 6x6. In 1987 a local body company refurbished this unit with a 350-gpm pump and an 850-gallon tank. The seal on the side of the tank is the State Seal. This unit is operated in conjunction with the State Forestry. *Photo by Mike Sanders.*

The Winters, California, Volunteer Fire Department operates this 1962 Chevy 60, ex CDF Model I wildland engine. It has a 375-gpm pump and a 500-gallon tank. The body builder is unknown. *Photo by Shaun P. Ryan.*

Ridge, New York, operates this 1963 International M-54 6x6 5-ton as a stump-jumper. This unit has a high-pressure pump and a 620-gallon tank. It is not unusual for fire departments on Long Island to operate several stump-jumpers. Ridge Fire Department operates several of these units. *Photo by Frank Wegloski.*

This 1963 International 4x4 has a 500-gpm front-mounted pump and a 250-gallon tank. The fire equipment was provided by American Fire Equipment. Richland Township, Michigan, Fire Department runs this unit. *Photo by Daniel A. Jasina.*

There are several volunteer fire departments in the city of New York. The Edgewater Park Volunteer Hose Company of the Bronx runs this 1964 Jeep FC with a Howe body. This unit has a 500-gpm pump and a 200-gallon tank. The Jeep FC was built from 1957 until 1965. *Photo by John M. Calderone.*

Todd Lincoln photographed this 1964 Chevy C30 4x4 pickup. The Barto, Pennsylvania, Fire Company operates Brush 22-5. It has a 60-gpm pump and a 150-gallon tank. Note the rear step on the back of the truck.

In the Heartland of America there are many poor rural communities; one such is Veale Township, Indiana. The volunteer fire department converted this 1964 International 4x4 into a firefighting unit. It has a 100-gpm pump and a 1400-gallon tank. It is not only used for wildland fires but for all fires. *Photo by Frank Wegloski.*

This M-35 2 1/2-ton 6x6 was a co-op truck given to the Silverton Fire Company of Dover Township, New Jersey. The New Jersey Forest Fire Service, in cooperation with the US Forest Service, obtains surplus military/government vehicles and transfers them to local municipal fire departments (all states have similar programs). This circa 1965 M-35 carried a 500-gallon water tank. It was replaced in the 1980s by a Dodge Power Ram 1-ton 4x4 chassis. The company turned this truck back over to the New Jersey Forest Fire Service, who then reassigned it to the Vincent Fire Company of Southamton Township. It served there until the early 1990s. *Photo by Scott Mattson, John Floyd Collection.*

In the 1950s and 1960s many fire companies in the East started using high-pressure for all kinds of firefighting. Originally developed by the Navy for shipboard firefighting, the John Bean Company marketed their high-pressure pumps and bodies starting in the 1940s. By the 1960s many fire apparatus came equipped with a high-pressure pump. This 1965 GMC was delivered to the Matawan Township (New Jersey) Hose and Chemical Co. (a.k.a. Oak Shades Fire Co.). This unit was used as an attack engine for structural fires, but was equipped with a winch in case the unit got stuck operating at a brush fire. The high-pressure pump could deliver more than 400 PSI out of the hose reel mount on top of the truck. *Photo by Bill Schwartz, Jim Burner Collection.*

Delran, New Jersey, Fire Co. #1 still runs this 1965 Dodge WM 300 Power Wagon that has been modified. This unit carries a 125-gpm pump and a 300-gallon tank. Brush 2316 is still in service. *Photo by Ron Jeffers.*

Quinebaug, Connecticut, had Farrar build the body on this 1965 Ford F-250 4x4 chassis. Engine 4 has a front-mounted pump and a 200-gallon tank. *Photo by Dick Bartlett.*

Bedford, New York, once owned this 1965 GMC/American 4x4 250-gpm unit. Now the Stevens Heights Engine Company #5 of the West Haven, Connecticut, Fire Department operates this unit, which is equipped with a 350-gallon tank and a front-mounted winch. *Photo by Frank Wegloski.*

Thornwood, New York, runs this 1965 GMC 4x4 pickup. Sanford mounted a 500-gpm pump and a 250-gallon tank on this unit. Note the lengths of hard suction hose mounted on the side. *Photo by Frank Wegloski.*

This rather long M-54 5-ton is in service with the Deer Park, New York, Fire Department. The make and year are unknown (circa 1965), but the unit has a 250-gpm pump and a 750-gallon tank. Deer Park also operates an M-35 2 1/2-ton stump-jumper and an M-35 2 1/2-ton tanker. *Photo by Robert P. Vaccaro.*

The Union Fire Company of Hamburg, Pennsylvania, operates this 1965 Kaiser M-35 2 1/2-ton 6x6. It was placed into service with the camouflage paint job. This truck is equipped with a 250-gpm pump mounted in front of the rear wheels on the driver side, and a 600-gallon tank. *Photo by Todd Lincoln.*

The state capital of Delaware is Dover. Dover has a volunteer fire department. This circa 1965 Dodge W-300 4x4 has been outfitted by American Fire Apparatus. It is believed to have a 150-gpm pump and a 200-gallon tank. Note the grill guard and the length of hard suction. *Photo by John D. Floyd.*

This 1965 Ford 4x4 chassis has a Young body. The unit has a 250-gpm pump and a 200-gallon tank. It was in service with the Nanjemoy, Maryland, Volunteer Fire Department. *Photo by Mike Defina Jr.*

Washington, Indiana, Fire Department has this 1965 International 4x4 with fire equipment by Midwest. This truck has a 750-gpm pump and a 300-gallon tank. Note the roller for the booster hose on the cab roof; this is done to protect the cab when stretching the booster hose over the cab to reach the front firefighting riding position. *Photo by Frank Wegloski.*

This 1965 Chevy 4x4 with a Midwest body has a 500-gpm pump and a 500-gallon tank. It is seen in service with the Haddon Township Fire Department of Carlisle, Indiana. Note the front firefighting riding position and enclosed hose reel. *Photo by Frank Wegloski.*

Sutphen outfitted the fire equipment on this 1965 GMC pickup for the Miami Township, Ohio, Fire Department. Field Unit 2 has a 250-gpm pump and a 200-gallon tank. *Photo by Greg Stapleton.*

The Griffith Park in Los Angeles, California, operates this 1965 GMC 5000 with a Yankee body. This engine has a 300-gpm pump and a 400-gallon tank. It was in service with the Los Angeles Fire Department before it was assigned to the Ranger Fire Patrol in Griffith Park. *Photo by Chuck Madderom.*

This 1965 Mack F tractor is hauling a 1978 Cat D6D. Transport and Dozer 6 are both in service with the Kern County, California, Fire Department at the Edison Station. *Photo by Chuck Madderom.*

This circa 1965 Bell Model 47 Helicopter was operated by a contractor to provide fire suppression to the New Jersey Forest Fire Service. It was in service until the late 1970s and the bucket could carry 100 gallons of water. *Photo by Maris Gabliks, NJ Forest Fire Service.*

The members of the New Jersey Forest Fire Service converted this 1966 Kaiser M-35 6x6 fuel tanker into a 1200-gallon fire service water tanker/tender. It has a 250-gpm pump and was placed into service in the year 2000. *Photo by John Rieth, NJ Forest Fire Service.*

The body on this 1966 International Loadstar 4x4 was manufactured in 1978 by Standard Welding. The Ellington, Connecticut, Fire Department operates Forestry 143. It carries a 250-gpm pump and a 400-gallon tank. *Photo by Glenn B. Vincent.*

Some may say this little 1966 Ford pickup is cute. The Long Hill, Connecticut, Fire Department had Maxim outfit this unit with a 250-gpm pump and a 150-gallon tank. *Photo by Glenn B. Vincent.*

The Cornplanter Township Fire Department of Oil City, Pennsylvania, runs this 1966 White M-35 2 1/2-ton 6x6. It is equipped with a 250-gpm pump and a 1000-gallon tank. *Photo by David Bowen.*

This is a Hahn brush truck; it is mounted on a 1966 Dodge W-300 chassis. The Upper Salford Township, Pennsylvania, Fire Company's Brush 78-41 has a 150-gpm pump and a 200-gallon tank. *Photo by Todd Lincoln.*

Rawlins County, Kansas, Fire & Rescue District #2 runs this 1966 International Fleetstar. The body was fabricated by the members of the Quinter, Kansas, Fire Department, before it was in service with Rawlins County. This unit has a 250-gpm pump and a 1000-gallon tank. This unit is not just a wildland unit but is a structural engine. This type of apparatus is common in the rural Midwest. *Photo by Dennis J. Maag.*

The New Jersey Forest Fire Service flies two 1966 UHH1 Huey helicopters. These aircraft carry a 300-gallon Bambi bucket that can be fitted under the aircraft. *Photo by John Rieth, NJ Forest Fire Service.*

After the Dodge M-37 4x4 the military needed a new truck, so Kaiser Jeep delivered the M-715 series 1 1/4-ton 4x4. The Kaiser Jeep Company produced these trucks from 1967-1969. This M-715 saw service with the Wells Corner Fire Company of Wells, Maine. Brush 2 has a 100-gpm pump and a 250-gallon tank. *Photo by Frank Wegloski.*

Dick Bartlett photographed this circa 1967 AMC M-715 ex military 4x4 that has been converted into a smaller sized brush-breaker. This unit is operated by the Carver, Massachusetts, Fire Department.

The Centerville-Osterville Fire District of Marstons Mills, Massachusetts, operated this rather unusual 1967 Maxim 6x6 brush-breaker. This unit has a 250-gpm pump and a 1000-gallon tank. Maxim was known to build brush-breakers for the Cape Cod area fire departments. *Photo Robert W. Fitz Jr.*

Haddam Neck, Connecticut, Fire Department operates this 1967 Chevy pickup that has a 10-gpm pump and a 75-gallon tank. *Photo Glenn B. Vincent.*

The snowplow shown on this 1967 M-715 Kaiser Jeep is not used for firefighting. Pawcatuck, Connecticut, Fire Department has equipped this unit with a 200-gpm pump, a 200-gallon tank, and a snowplow. Many fire companies have equipped brush units with snowplows for clearing snow at the fire station or if need be, to escort the fire equipment through the snow. *Photo by Frank Wegloski.*

The bumper on this 1967 Kaiser Jeep M-715 4x4 has a large brush guard to push away brush. This unit, operated by the Avon, Connecticut, Fire Department, has a 250-gpm pump and a 300-gallon tank. *Photo by Glenn B. Vincent.*

Not all wildland firefighting apparatus on Long Island are stump-jumpers! This 1967 Ford F-700 4x4 equipped with a 750-gpm front-mounted pump, is in service with Medford, New York, Fire Department. The small New York fire apparatus manufacturer known as Young built the body. *Photo by Frank Wegloski.*

The Pennsylvania Department of Conservation & Natural Resources Bureau of Forestry has this 1967 Mercedes Benz Unimog 4x4. This unit is operated by the Waterville, Pennsylvania, Fire Department. This unit has a small pump and a 250-gallon tank. *Photo by David Bowen.*

In Littlestown, Pennsylvania, the Alpha Fire Department runs this well kept 1967 GMC 4x4 300-gpm brush truck. It has a Bruco body with a 300-gallon tank. *Photo by Mike Sanders.*

The National Zoological Park, Front Royal, Virginia, Fire Department runs this 1967 Kaiser M-715 4x4. It has a 115-gpm pump and a 300-gallon tank. It is nicknamed "Lil' Bucket." *Photo by Mike Sanders.*

When we said that some rural places of the US were very poor it was not a joke! The Martin County, Indiana, Civil Defense operates this 1967 International Meyers, old sewer jet truck for fire suppression. It has a high-pressure pump and a 1000-gallon tank. *Photo by Frank Wegloski.*

This circa 1968 Dodge W-500 4x4 chassis has one of the 1940 bodies that were mounted on the 1940s Chevys (page 6). The Dodge W-500 series was the heaviest in the Power Wagon line. *Photo by Maris Gabliks, NJ Forest Fire Service.*

The Massachusetts Department of Forestry operates several brush-breakers on and near Cape Cod. This 1968 International Loadstar 4x4 was outfitted by a small fire apparatus company called Farrar. This unit has a 250-gpm pump and a 650-gallon tank. *Photo by Robert W. Fitz Jr.*

This unusual 1968 Chevy chassis has a Progress body. West Greenwich Fire Company No. 1 of Nooseneck Hill, Road Island, has in service Brush 24. It has a 500-gpm pump and a 500-gallon tank. *Photo by Glenn B. Vincent.*

There are several companies that manufacture stump-jumpers on Long Island, and each has their own style of fabricating the brush bars. This 1968 Reo M-35 6x6 2 1/2-ton is in service with the East Hampton Fire Department. *Photo by Frank Wegloski.*

Fire departments can be very creative when constructing wildland fire apparatus. The Aultman, Pennsylvania, Fire Department had this modified 1968 Chevy Suburban 4x4 converted into a fire apparatus. It has a 250-gpm pump and a 250-gallon tank. Also, hard suction hose and a ladder are carried on the roof. *Photo by David Bowen.*

Bendersville, Pennsylvania, Volunteer Fire Company members converted this 1968 Kaiser M-54 5-ton 6x6 into a brush truck in 1997. This rig carries a 250-gpm pump and a 500-gallon tank. Note the front firefighting riding position. *Photo by Howard Meile.*

The Liverpool, Pennsylvania, Fire Company runs this attractive 1968 Kaiser M-715 4x4. This unit has a 100-gpm pump and a 275-gallon tank. *Photo by Todd Lincoln.*

This 1968 International 4x4 with a Hahn body did serve the Minquadle, Delaware, Fire Department. Now, this unit, equipped with a 250-gpm pump and a 300-gallon tank, is in service with the Reva, Virginia, Volunteer Fire & Rescue. *Photo by Mike Sanders.*

Howard Meile photographed the Mt. Airy, Maryland, Volunteer Fire Company's 1968 Kaiser M-715 4x4. Brush 15 has a 90-gpm pump and a 200-gallon tank. Note the extensive brush protection. It has the nickname of "Pound Puppy," and is equipped with a Mack Bulldog emblem on the hood.

Dennis J. Maag photographed this 1968 International Scout 4x4. It has a 250-gpm pump and a 150-gallon tank. It is operated by the Wentzville, Missouri, Volunteer Fire Department.

This 1968 GMC 4500 is in service with the Redwood Valley Calpella, California, Fire Department. It has a 600-gpm pump, a 750-gallon water tank, and a body built by the CDF. *Photo by Stefan Farage.*

Evergreen, Colorado, operates several water tenders with brush protection. This is Tender 74, a 1968 International 4x4 with a locally built body. It has a 750-gpm front-mounted pump and a 1500-gallon tank. *Photo by Rick Davis.*

The New Jersey Forest Fire Service flies this 1968 Bell Ranger. It is used for observation. A small 100-gallon bucket can be placed under the aircraft if needed. *Photo by John Rieth.*

This late 1960s FWD 6x6 was operated by the Massachusetts Forestry Service. This is another photo where little information is known. Note the heavy brush guards. *Photo by Frank Wegloski.*

During the later part of the 1960s the New Jersey Forest Fire Service began to use several Dodge W-300 4x4s as brush trucks. This late 1960s W-300 had a 200-gpm pump and a 300-gallon tank. *Photo by Bill Schwartz, Jim Burner Collection.*

This circa 1969 Ford 4x4 was built by the local fire apparatus company called Cherefko and Son of Old Bridge, New Jersey. This unit had a 150-gpm pump and a 200-gallon tank. Cherefko built a small number of brush units, small pumpers, and tankers for mostly New Jersey Fire Departments. This unit was operated by the Plainsboro, New Jersey, Fire Co. until the 1990s. *Photo by Scott Mattson.*

Island Heights, New Jersey, Fire Company still operates this 1969 M-52 series 5-ton Kaiser tractor, pulling a 6000-gallon tank trailer. This tanker (tenders are still called tankers in the East) is not only used to supply water for wildland firefighting operations, but for all types of fires. There is no pump—a pumper must draft from the tank. *Photo by John Rowe.*

This 1969 Jeep Gladiator was in service with the Glastonbury, Connecticut, Fire Department. Forestry 44 comes equipped with a 150-gpm pump and a 100-gallon tank. The Jeep Gladiator was first produced in 1963; the M-715 was based on this truck. By 1987 Jeep stopped production of the full-sized pickup. The J-10/J-20 were the two models of pickups in the Gladiator line. *Glenn B. Vincent Photo.*

This 1969 Dodge W-300 4x4 flatbed, called the "Hose Draggin' Power Wagon," is in service with the Suedburg, Pennsylvania, Fire Department. It has a 50-gpm pump and a 100-gallon tank. *Photo by David Bowen.*

This monster truck-like field piece is a 1969 GMC 4x4 with a lift kit and very large tires. It was in service with the Port Penn, Delaware, Volunteer Fire Department. It is now owned by a collector and was used to help fight fires in a marsh area. *Photo by Scott Mattson.*

Maryland Department of Natural Resources had this 1969 Dodge 600 rollback with a 1966 John Deere 350 dozer with a fire plow in service as Eastern 53. *Photo by Howard Meile.*

The New Jersey Forest Fire Service converted this 1970 CDE Gamgoat M-561 6x6. This vehicle is an amphibious articulated all-terrain vehicle. This unit is stationed in the hilly northwestern part of New Jersey and is fitted with a 250-gpm pump and a 300-gallon tank. The M-561 was built by Consolidated Electric Co. from 1969-1977. *Photo by Adam Alberti.*

Dick Bartlett photographed this M-54 6x6 5-ton that has been converted into a brush-breaker. The Plainville, Massachusetts, Fire Department operates this circa 1970 unit. Note the large brush bars in the front; the bars force trees to be pushed in front of the trucks to avoid damaging the truck.

The Connecticut Forest Fire Service operates this circa 1970 M-35 2 1/2-ton 6x6. This unit carries a large tank mounted in the cargo bed. *Photo by Alan Mudry.*

This stump-jumper was manufactured by Long Island Brush Trucks, who have outfitted hundreds of such units over the years. This 1970 Mack M-54 equipped with a 500-gpm pump and a 1400-gallon tank was delivered to the Farmingville, New York, Fire Department in 1999. *Photo by Robert P. Vaccaro.*

This unit is used for fighting fires and rescuing people trapped by floods. It was built using ideas from stump-jumpers. In 1995 the Lindenhurst, New York, Fire Department had this 1970 Mack M-54 5-ton converted into a flood unit by Chavis. The "Storm Fighter" sports brush protection to keep fallen or low branches from damaging the truck. Also, the push pads on the bumper are like a push boat's bow, made to push away underwater obstructions, like cars! This unit has a 250-gpm pump and a 300-gallon tank. A hole was cut in the cargo bed near the pump to allow a suction hose to be lowered under the truck to provide a water supply. *Photo by Robert P. Vaccaro.*

The Packard Company of the Bridgehampton, New York, Fire Department, runs this smaller version of a stump-jumper. This 1970 International 4x4 Loadstar has a 500-gpm pump and a 500-gallon tank. *Photo by Frank Wegloski.*

This 1970 International 4x4 was refurbished by Pro-Tech in 1992. The Community Fire Company of Virginville, Pennsylvania, has Brush 33-5 in service equipped with a 300-gpm pump and a 320-gallon tank. This rig was in service with the Newtown Square, Pennsylvania, Fire Department, before serving with Community Fire Company. *Photo by Todd Lincoln.*

In Boyertown, Pennsylvania, the Friendship Hook & Ladder Co. runs this 1970 Dodge W-300 4x4. It has an Allegheny body that was refurbished in 1997. This maroon-colored truck has a 250-gpm pump and a 200-gallon tank. *Photo by Todd Lincoln.*

Brady Township, Pennsylvania, Fire Department runs this 1970 Kaiser M-35 6x6 as Brush Tanker 30. This unit has a 650-gpm pump and an 1100-gallon tank. *Photo by Patrick Shoop.*

On the Outer Banks of North Carolina the Nags Head Fire Rescue operates this circa 1970 Kaiser M-35 2 1/2-ton 6x6. It has a 150-gpm pump and a 300-gallon tank. *Photo by John Rieth.*

Comstock, Michigan, Fire Rescue runs this 1970 Ford 4x4 pickup. It carries a slip-on pump and tank. The pump is rated at 200-gpm and the tank holds 150 gallons. *Photo by Daniel A. Jasina.*

The CDF in Ukiah, California, runs this 1970 Ford L9000 with a 1982 Cat D-4E dozer. *Photo by Shaun P. Ryan.*

Nevada Division of Forestry at Galena Creek ran this 1970 International with a Douglas body. It carried a 125-gpm pump and a 400-gallon tank. *Photo by Garry Kadzielawski.*

Los Angeles County Fire Department Helicopter 15 is a 1970 Bell 205 fitted with a water tank under the aircraft. As of spring 2001, LA County flies eight helicopters—one Bell 206 Jet Ranger, three Bell 205 helicopters, and four Bell 412 helicopters. In the summer of 2001, two new Sikorsky S-70A Firehawks, which will each have 1,000-gallon tanks, will be placed into service. The Firehawk is the firefighting model of the US Military's Blackhawk. *Photo by Robert P. Vaccaro.*

The Nevada Division of Forestry had this early 1970s Kaiser M-813 6x6 rebuilt in 1999 by Burton Fire Apparatus, with a Yuba Tank. It has a 500-gpm pump and a 1550-gallon tank. It is stationed in Washoe County, Nevada. *Photo by Shaun P. Ryan.*

The Florida Department of Forestry operates this former Army Bell AH-1P Cobra circa 1970 attack helicopter. They are now called Fire Snakes. Its only weapon now consists of a 320-gallon Bambi bucket. Florida Division of Forestry flies four Super Hueys, three Fire Snakes, and two OH-58 helicopters, in addition to 19 fixed wing aircraft. *Photo by Joe Abrams.*

This 1971 International 4x4/Bruco brush unit was used by the Weekstown Fire Co. of Mullica Township, New Jersey. This truck has a 150-gpm pump and a 500-gallon tank. Bruco built many brush units, many which had diamond-plated compartments. The diamond-plates helped keep a good appearance when the truck was exposed to constant scratches from fighting brush fires. *Photo by Scott Mattson.*

Hahn Fire Apparatus, which was a favorite builder in the New Jersey area, had built several brush units. This 1971 International Loadstar chassis had a 300-gpm pump and a 500-gallon tank. This 4x4 unit was built by Hahn for the Hamilton Fire Company of Neptune, New Jersey. *Photo by Scott Mattson.*

Kezar Falls, Maine, runs this interesting 1971 International Loadstar 4x4 2000-gallon tanker. This unit has a 250-gpm pump and a front-mounted winch. *Photo by Frank Wegloski.*

Some fire departments in Maine refer to their brush trucks as tanks. This 1971 Ford F-700 brush unit is in service with the Brunswick, Maine, Fire Department. Tank 1 has a Howe body, a 100-gpm pump, a 650-gallon water tank, a 50-gallon foam tank and a front brush guard. *Photo by Frank Wegloski.*

Kreamer, Pennsylvania, Fire Company runs this 1971 Chevy C10 4x4 flatbed. Brush 12-1 has a 500-gpm pump and a 300-gallon tank. *Photo by Todd Lincoln.*

Greg Stapleton photographed and reported that Burgin, Kentucky, Tanker 655 is a 1971 Kaiser 530-C pumper. It now has a 750-gpm pump and a 1200-gallon tank. The 530-C military tactical fire apparatus had a foam gun mounted behind the cab. Such manufacturers as Fire Master, Ward LaFrance, and Engineered Devices Inc. produced both the 530-B and 530-C.

In central and southern New Jersey there were so many Dodge Power Wagon brush trucks, that in some areas the name Power Wagon is referred to as a brush truck, no matter who really manufactured the truck. Here, the Waterford Fire Company of Winslow Township operated this 1971 Dodge Power Wagon. On the fender the unit is lettered Power Wagon 2585. The body builder is Welch. The PTO-driven pump is rated at 350 gpm. This tank holds 300 gallons. *Photo by Scott Mattson.*

The military had another batch of tactical fire apparatus built from the 1950s to the early 1970s. The 530 series pumpers were based on the M-44 chassis, which is a variant of the M-35 2 1/2-ton 6x6. The first model was the 530-A equipped with a front-mounted pump, then came the 530-B which had a midship-mounted pump. The last model was the 530-C which came equipped with foam equipment, including a roof-mounted foam turret. The fire apparatus was manufactured by such companies as Dakota, Hesse, Howe, FireMaster, Ward LaFrance, and Engineered Devices Inc. The Pine Beach Fire Co., New Jersey, received this 1972 530-B, which saw service with the NJ Air National Guard. It is equipped with a 500-gpm pump and a 400-gallon tank. *Photo By John Rowe.*

Pine Rock Park Volunteer Fire Company of Shelton, Connecticut, once operated this 1972 International 4x4. The body had a 275-gpm pump and a 200-gallon tank and was manufactured by Imperial. Imperial was started in Rancocas, New Jersey, in the early 1970s, by people who used to work at Hahn. This truck was once operated by the North Branford, Connecticut, Fire Co. #1. *Photo by Frank Wegloski.*

Cooperstown, Pennsylvania, Fire Department runs this 1972 AM General (Kaiser) 530-B pumper. It was later refurbished by New Lexington. This rig has a 500-gpm pump and a 500-gallon tank. *Photo by David Bowen.*

Washington Township Fire Department of Fryburg, Pennsylvania, runs this 1972 International Loadstar equipped with a 350-gpm pump and a 600-gallon tank. Fire department members converted this chassis into a wildland firefighting apparatus. *Photo by David Bowen.*

In America there is freedom to use any type of vehicle to convert into a wildland firefighting unit. Here we see that Glyndon, Maryland, Volunteer Fire Department, used their freedom to convert this 1972 International Scout SUV into a brush truck. It has a 150-gpm pump and a small 68-gallon tank. *Photo by Frank Wegloski.*

The Maryland Department of Natural Resources operated this 1972 Mercedes Benz Unimog 4x4. This unit had a high-pressure pump and a 250-gallon tank. *Photo by Howard Meile.*

United Communities Fire Department of Romancoke, Maryland, runs this 1972 AM General 530-B ex military pumper as a brush truck. It has a 500-gpm pump and a 400-gallon tank. *Photo by Howard Meile.*

French Village Fire Department of Fairview Heights, Illinois, has this 1972 GMC 4x4 pickup in service with fire equipment by Tower Fire Apparatus. Brush 254 has a 250-gpm pump and a 100-gallon tank. It was refurbished later by Midwest. *Photo by Dennis J. Maag.*

In East Olympia, Washington, the Thurston County Fire District 6 has this 1972 AM General M-813 5-ton 6x6 in service. This unit has a 1993 H&W body that is equipped with a 500-gpm pump and a 1600-gallon tank. *Photo by Glenn B. Vincent.*

The New Jersey Forest Fire Service operates this 1972 International Loadstar chassis, which has a body that has been used on at least two older chassis dating back to the 1940s. A small 250-gpm pump and a 500-gallon tank provide the firefighting package for this unit. In addition, this truck tows a 1977 John Deere 350 dozer with a fire line plow. *Photo by John Rieth, NJ Forest Fire Service.*

Throughout America, even large urban areas have some areas of open space or undeveloped areas, or even nature preserve/parks. To help fight fires in these areas, large urban fire departments have converted older apparatus or have purchased new apparatus to fight wildfires. The new regional North Hudson, New Jersey, Fire Department converted this ex Union City, New Jersey, Fire Department Engine 2, a 1973 Seagrave, into a brush fire unit. This unit has a 1500-gpm pump and a 525-gallon tank. *Photo by Ron Jeffers.*

The Jackson Township Fire Department of Mineral Point, Pennsylvania, runs this nice 1973 International 4x4 Loadstar with a John Bean body. The unit has a high-pressure pump and a 375-gallon tank. *Photo by David Bowen.*

Todd Lincoln shot this 1973 Jeep J2000 full-sized pickup. Earl Township, Pennsylvania, Fire Company uses Brush 19-5; it has a 250-gpm pump and a 200-gallon tank.

This 1973 International Pierce mini-pumper used to be in service with Hulmeville, Pennsylvania, Fire Department. This brush truck is now operated by the Community Fire Company of New Ringgold, Pennsylvania. It was refurbished in 1986 and carries a 300-gpm pump and a 250-gallon tank. *Photo by Todd Lincoln.*

This 1973 AM General M-35 6x6 has an unusual body built by Davies. Tanker 28 is in service with the Glendale Volunteer Fire Department of Coalport, Pennsylvania. *Photo by Patrick Shoop.*

The Union County West End Fire Company of Glen Iron, Pennsylvania, has this 1973 AM General M-813 5-ton 6x6 in service. It has a 328-gpm pump, a 1000-gallon tank and a collapsible portable tank. *Photo by Todd Lincoln.*

This 4-door 1973 International 4x4 pickup has fire equipment by Howe. This St. Joseph Township, Indiana, Fire Department's truck has a 35-gpm pump and a 200-gallon tank. *Photo by Frank Wegloski.*

In 1987 The Nevada Division of Forestry had a Pauls 3200-gallon tank placed onto this 1973 Ford W Series Cab-Over Chassis. It has a 250-gpm pump. *Photo by Garry Kadzielawski.*

The Laramie County, Wyoming, Fire District #2 uses this 1973 AM General M-35 2 1/2-ton 6x6 with a locally built body. This rig has a 300-gpm pump, a 750-gallon water tank, and a 20-gallon tank of class A foam. *Photo by Rick Davis.*

The New Jersey Forest Fire Service runs this 1973 Cat D7 dozer. *Photo by John Rieth.*

The Prince Frederick, Maryland, Volunteer Fire Department runs this little 1974 Jeep 4x4. It carries a 60-gpm pump and an 80-gallon tank. *Photo by Frank Wegloski.*

The Ontario, California, Fire Department, uses this 1974 International Paystar 4x4 with a Van Pelt body as Water Tender 137. It has a 400-gpm pump and a 1000-gallon tank. *Photo by Chuck Madderom.*

The military also uses commercial trucks mainly to support base functions. This circa 1975 International Loadstar 4x4 flatbed was operated by the army at nearby Fort Monmouth before the unit was obtained by the Eatontown Fire Department. The volunteers converted the truck into a brush truck equipped with a 400-gallon tank and a 300-gallon tank. A small 200-gpm pump was also fitted. *Photo by John Rowe.*

Originally delivered to the Transboro, New Jersey, Fire Company, this circa 1975 Ford F-800 4x4 Bean apparatus did not have the brush protection. During the early 1990s the unit was refurbished by G&S. This unit carries the famous John Bean high-pressure pump and a 500-gallon water tank. *Photo by Scott Mattson.*

This ex Air Force 1975 Dodge CT 800 model 5000-gallon fuel tanker still provides aircraft refueling to the New Jersey Forest Fire Service at the Colye Field Air Base in Woodland Township, New Jersey. The tank was built by the Consolidated Diesel Electric Company. *Photo by John Rieth, NJ Forest Fire Service.*

Hamlin, Pennsylvania, Fire Rescue operates two M-54 5-ton 6x6 trucks. This one is a 1975 Mack set up for brush fire attack. It has a 350-gpm pump, two 500-gallon military tanks set in the cargo body, and is painted blue and white in a camouflage paint scheme. *Photo by John M. Calderone.*

Hank Knight photographed the St. James, Tennessee, Volunteer Fire Department's 1975 Ford pickup. This unit has a 300-gpm pump and a 400-gallon tank.

All gray is the color of this 1975 AM General M-54 5-ton 6x6. The Elmore Township, Indiana, Fire Department unit has a 2300-gallon tank. *Photo by Frank Wegloski.*

The Jefferson R-7 Fire Protection District of Jefferson County, Missouri, has this 1975 International 4x4 with a SCAT body. It has a 350-gpm pump, a 200-gallon tank, and carries five gallons of class A foam. *Photo by Dennis J. Maag.*

Waterloo, Illinois, Rural Unit operates this 1975 Chevy 4x4 Towers as Mud Buggy 281. It has a 250-gpm PTO pump and a 250-gallon tank. Most wildland fire apparatus have either a PTO-driven pump (power take off unit from the vehicle's engine), or a pump that is powered by a separate engine. *Photo by Dennis J. Maag.*

Anything goes for wildland fire apparatus. Here we see a 1975 Lockheed 8x8 truck with B&Z fire equipment that was built for the US Bureau of Land Management in Carson City, Nevada. This unit has a 100-gpm pump and a 1050-gallon tank. *Photo by Chuck Madderom.*

Otis Air Force Base, Massachusetts, operates several brush-breakers. This 1976 AMC M-54 has a Gibson body. In 1995 the unit was rebuilt, and it now carries a 300-gpm pump and a 950-gallon tank. *Photo by Robert W. Fitz Jr.*

This 1976 Dodge 4x4 chassis was an ex military truck, on which a 1991 Omaha utility body was mounted. The unit, operated by the Sandy Hook, Connecticut, Fire Department, has a 250-gpm pump and a 250-gallon tank. *Photo by Glenn B. Vincent.*

The Maryland Department of Natural Resources runs this 1976 GMC/Brown-Minneapolis tanker as Eastern 62. This water tender has a 1000-gpm pump and a 1250-gallon tank. *Photo by Howard Meile.*

Jasonville, Indiana, Fire Department uses this 1976 Chevy with a local body. It has a 300-gpm pump and a 500-gallon tank. *Photo by Frank Wegloski.*

In 1976 the military received a large number of Dodge 1 1/4-ton 4x4s. These were based on the civilian Dodge pickup of the time. Known as the 880 series, more than eight variants were produced from 1976-1977. This 1977 M-880 was converted into a brush fire truck in the early 1990s by the Earle Naval Weapons Station Fire Department at Earle, New Jersey. Later, the unit was obtained by the New Jersey Forest Fire Service and now serves at Cheesequake State Park. The unit is equipped with a 200-gpm pump and a 200-gallon tank. *Photo by John Rieth, NJ Forest Fire Service.*

Squankium Fire company of Howell, New Jersey, received this 1977 AMC M-35A2 from the New Jersey Forest Fire Service through a cooperative agreement in 1999. This unit carries a compressed air foam system (CAFS), a 500-gallon water tank and a 200-gpm pump. The brush protection was fabricated by Megills Inc., of Howell, New Jersey. *Photo by John Rieth.*

Mark A. Redman photographed this 1977 Dodge pickup. The members of the Glastonbury, Connecticut, Fire Department outfitted this truck themselves with a 250-gpm pump, a 150-gallon tank, and brush protection in 1996. This unit replaced the 1969 Jeep Gladiator seen on page 64.

Cornish, Maine, Fire Department uses this 1977 International Loadstar 4x4. The body is typical of a military cargo truck, thus at least the body may have been ex military. This unit has a 500-gpm front-mounted pump and an 1100-gallon tank. *Photo by Frank Wegloski.*

Madera, Pennsylvania, Fire Department has this 1977 Dodge military M-880 4x4. It has been modified with civilian grill and sun visor. It has a slip-on pump and tank. *Photo by Patrick Shoop.*

Mike Sanders photographed this 1977 Ford Bronco 4x4 that has been converted into a wildland fire apparatus. The Dumfries-Triangle Vol. Fire Department's Brush-17 comes equipped with a 35-gpm pump and an 80-gallon tank. Note the hoops running around the wheel wells; they keep branches or limbs from getting caught in the wheel wells.

Mike Sanders not only photographed this 1977 Jeep CJ7, he operated this truck. His fire department of Ashburn, Virginia, in Loundon County, ran this truck equipped with brush protection, a 65-gpm pump and a 35-gallon tank until it was sold in 2000. Jeep 6 used to serve the Sterling, Virginia, Fire Department before its stay in Ashburn.

Daniel Jasina photographed Richland Township, Michigan, Fire Department's 1977 Dodge M-880 4x4. It has a 250-gpm pump, a 150-gallon tank, and a spray bar on the front bumper.

This unlettered wildland engine belongs to the Busseron Township Fire Department of Toaktown, Indiana. T-13 is a 1977 International Loadstar 4x4 with a Quality body. The pump is rated at 300-gpm and the tank can hold 1000 gallons. This truck saw service with the Vincennes Township Fire Department prior to Busseron Township. *Photo by Frank Wegloski.*

This retired California Department of Forestry 1977 International with a B&Z body is now in service with the Storey County, Nevada, Fire Department. Brush 1 has a 500-gpm pump and a 500-gallon tank. *Photo by Garry Kadzielawski.*

The Brevent Park and Leonardo Fire Co. of Middletown, New Jersey, received this 1978 Dodge W-200 pickup in 1997 from the Point Pleasant, New Jersey, Fire Co. #1, who in turn bought it from the Laurlton Fire Co. in Brick, New Jersey. This unit was based on the New Jersey Forest Fire Service units at the time. It has a Darley 200-gpm pump and 240-gallon tank. *Photo by John Rieth.*

Some people believe they must cage up their pets. The Gorden's Corner Fire Co. of Manalapan, New Jersey, had caged their 1978 Dodge/Pierce mini-pumper. The unit was also fitted with a spray bar ground sweeps in the front of the bumper. The Gorden's Corner Fire Company retrofitted this unit into this configuration during the late 1990s. It was equipped with a 300-gpm pump and a 250-gallon tank. *Photo by John Rieth.*

This Arcadia, Maryland, 1978 Dodge pickup's body was modified by Page Lambert. It sports a 300-gpm pump and a 200-gallon tank. Note the snowplow attachment in the front and the pump panel cut in the pickup body. *Photo by Frank Wegloski.*

Brush 55 of the Weyers Cave, Virginia, Volunteer Fire Company had this 1978 Ford F-250 4x4 refurbished by M&W in 1985. It has a 225-gpm pump and a 200-gallon tank. *Photo by Mike Sanders.*

Battletown, Kentucky, Fire Department uses this 1978 Ford F-700 4x4 with a Warner utility body. It has a 350-gpm pump and a 600-gallon tank. *Photo by Greg Stapleton.*

San Diego, California, operated a 1978 International La Mesa wildland engine. It has a 420-gpm pump and 600-gallon tank. *Photo by Frank Wegloski.*

This ex Air Force fuel truck now serves the Nevada Division of Forestry. It is assigned to the Spring Creek, Nevada, Volunteer Fire Department in Elko, Nevada. This 1978 White with a Macleod body was refurbished in 1991 by the Nevada Division of Forestry. This unit has a 100-gpm pump and a 2600-gallon tank. *Photo by Garry Kadzielawski.*

This 1978 John Deere 350 wide track dozer is fitted with a 150-gpm pump and a 200-gallon tank. It was made to go through swamps and other wet areas. It is in service with the NJ Forest Fire Service. *Photo by John Rieth, NJ Forest Fire Service.*

The Leicester, Massachusetts, Fire Department operates this "Forestry Tanker." It is a 1979 International Loadstar 4x4 with a body built by Farrar. A 250-gpm pump is connected to a 1000-gallon tank. *Photo by Dick Bartlett.*

Mount Penn, Pennsylvania, runs this 1979 GMC 4x4 with a locally built body. This unit has a portable pump hooked up to a 150-gallon water tank. *Photo by Todd Lincoln.*

Brush 92 of the Osceola County, Florida, Fire Rescue is a 1979 GMC 4x4/E-ONE 250-gpm rig. It has a 200-gallon tank and saw service with Reedy Creek Fire Rescue at Disney World before its current home. *Photo by Glenn B. Vincent.*

Jefferson Fire Protection District of Mt. Vernon, Illinois, has this 1979 Jeep in service. It has a 20-gpm pump and a 60-gallon tank. *Photo by Frank Wegloski.*

Bristol, Wisconsin's Fire Department uses this 1979 GMC 4x4 equipped with a slip-on pump and tank. Note the length of hard suction hose above the toolbox on the driver's side. *Photo by Mike Tenerelli.*

The CDF in Temecula, California, has a 1979 International Cargostar with a Westates wildland engine body. It has a 500-gpm pump and a 650-gallon tank. *Photo by Chuck Madderom.*

US Forest Service Engine 37 at San Bernardino, California, National Forest runs this 1979 International B&Z wildland engine. It has a 275-gpm pump and a 500-gallon tank. *Photo by Chuck Madderom.*

New Jersey Forest Fire Service operates more than 22 John Deere 350, or 450 bulldozers. This 1979 JD 350 has a fire line plow on the rear. *Photo by John Rieth, New Jersey Forest Fire Service.*

A 1980 International Cargostar truck tractor that saw service with the US Air Force was obtained by the New Jersey Forest Fire Service in the mid-1990s. The truck was converted into a transporter for a 1977 John Deere 350 dozer/plow unit. *Photo by John Rieth, NJ Forest Fire Service.*

This small airport crash truck is now used for wildland firefighting in the New Jersey Meadowlands, which encompasses thousands of acres of marshlands near the Giants Stadium. Moonachie, New Jersey, Fire Department's Engine 9 is a 1980 International/Gibson 750-gpm crash truck. This unit, fitted with a 1000-gallon water tank and a 150-gallon foam tank, used to be operated by the Teterboro Airport Fire Department. *Photo by Ron Jeffers.*

Hawthorne, New York, Fire Department has this 1980 Chevy 4x4 with an E-ONE body in service. This unit has a 350-gpm pump and a 300-gallon tank. *Photo by Frank Wegloski.*

The City of Reading, Pennsylvania, Bureau of Fire, has both career and volunteer firefighters. Brush 1, a 1980 Chevy C30 4x4, has a Car Mar body equipped with a 350-gpm pump and a 250-gallon tank. It was assigned to the Union Fire Company. *Photo by Todd Lincoln.*

The Naval Air Station in Cecil Field, Florida, operated this 1980 AM General M-52 (the tractor version of the M-54), pulling a 1983 FMC Fire Trac vehicle (see page 120). After Cecil Field closed, this unit was moved to Naval Station Mayport, Florida. *Photo by Mark V. Carr.*

In the Midwest, brush cages are not very popular. This 1980 Chevy Steelweld is an exception. The St. Clair, Missouri, Fire Protection District has Brush 321 in service equipped with a 250-gpm pump, a 250-gallon tank, and a brush cage. *Photo by Dennis J. Maag.*

Garry Kadzielawski photographed this 1980 GMC Brigadier tractor hauling a 1973 Cat dozer. Heavy Transport 4442 is operated by the CDF at Angels Camp, California.

This circa 1980 Grumman Agcat biplane is a crop duster, however, many states contract such aircraft for wildfire suppression duty. The New Jersey Forest Fire Service contracted this and nine more Agcat planes from the Downstown Airocrop Corp. of Downstown, New Jersey. These aircraft can hold up to 300 gallons of water or water-class A foam mixture. *Photo by John Rieth, NJ Forest Fire Service.*

This all-black 1981 GMC General tractor was acquired in the early 1990s by the Minotola Volunteer Fire Company in Buena, New Jersey. A 1965 Heil 9000-gallon tank trailer was converted from a fuel tanker into a water tanker. This is the largest tanker in New Jersey. It provides water not just for wildfire fighting, but also to any emergency requiring a mobile water supply. *Photo by Scott Mattson.*

The Massachusetts Department of Environmental Management, Bureau of Forest Fire Control, operates this circa 1981 GMC Astro tractor, pulling this Smokey Bear display trailer. This tractor can also pull lowboy trailers with heavy equipment. *Photo by Dick Bartlett.*

The Friendship Hose Co. #1 of Newville, Pennsylvania, has this 4-door 1981 Dodge 4x4 in service. It has a Stahl utility body, a 150-gpm pump and a 200-gallon tank. Note the front firefighting riding position. *Photo by Todd Lincoln.*

Some fire departments prefer a pickup style truck instead of a flatbed or utility body—there may be financial reasons for this as well. The Morgan Fire Co. of Sayreville, New Jersey, received this 1982 Chevy 4x4 pickup. A local welder fitted the unit with a brush cage. This truck has a 200-gpm pump and a 200-gallon tank. *Photo by John Rieth.*

Jeeps in the CJ series have been used by the fire service from the end of World War II right up to today. This 1982 Jeep CJ is operated by the North Centerville Vol. Fire Company in Hazlet, New Jersey. It has a 150-gpm pump and a 100-gallon tank. *Photo by John Rieth.*

On the Outer Banks of North Carolina the Duck Volunteer Fire Department got a 1982 International 4x4 with a utility body on it donated by the Carolina Power Company. The members converted the truck into a brush truck. It has a 250-gpm pump and a 400-gallon tank. *Photo by John Rieth.*

Also in service with the Hooksett, New Hampshire, Fire Department is this 1982 Chevy/Saulsbury 4x4 mini-pumper. Forestry 2 has a 300-gpm pump and a 200-gallon tank. *Photo by Glenn B. Vincent.*

Private for profit fire suppression companies are not heard of in the East, however many private contractors provide structural and wildland engines, water tenders (tankers), heavy earthmoving equipment, aircraft, and even firefighters for firefighting in western states. Here we see the Western Heavy Equipment Company's 1982 Freightliner with a 1979 Cozad lowboy trailer hauling a 1992 Cat D6H dozer. The contractor's equipment can be seen at any fire throughout the western states. This truck and dozer finds Escondido, California, as home base. *Photo by Chuck Madderom.*

This rig is another contractor's apparatus. This 1982 Ford LN 9000 has a Denver Truck Sales body. It has a 750-gpm pump and a 2000-gallon tank. It is owned and operated by Western Heavy Equipment Co. of Escondido, California. *Photo by Chuck Madderom.*

The CDF in Perris, California, runs this 1982 International with a Weststates body. This wildland engine has a 300-gpm pump and a 500-gallon tank. Note the style of the body; hundreds of CDF units share the same style body. *Photo by Chuck Madderom.*

This attractive 1982 International 6x6 is in service as Water Tender 238 of the San Ramon Valley, California, Fire Protection District. It has a Van Pelt body equipped with a 333-gpm pump and a 1500-gallon tank. *Photo by Shaun P. Ryan.*

This 1982 Mack Superliner normally has a trailer hauling a dozer with it, however, this 6000-gallon tank trailer is pulled by a dozer transport tractor when needed. This tanker supplies water to large wildfires, large base camps where thousands of firefighters are camping when fighting a prolonged campaign wildfire, or in any other emergency needing a large supply of water. This unit is operated by Los Angeles County Fire Department. *Photo by Chuck Madderom.*

Los Angeles County, California, Fire Department runs this 1982 Mack Superliner as Transport 7. It hauls a 1965 Cat D8H dozer. It is stationed in Lancaster, California. *Photo by Chuck Madderom.*

This unusual 1982 FWD 4x4 has a body built by the members of the San Bernardino, California, Fire Department. Water Tender 3 has a 500-gpm pump and a 2250-gallon tank. *Photo by Chuck Madderom.*

Honolulu, Hawaii, has in service this ex-CDF 1982 International with a Weststates body. This unit has a 300-gpm pump and a 500-gallon tank. *Photo by Shaun P. Ryan.*

Gowan-Knight, a small fire apparatus builder/re-builder from Connecticut, built this 1983 GMC 4x4 brush unit for Simsbury, Connecticut. This unit has a 250-gpm pump and a 250-gallon tank. *Photo by Glenn B. Vincent.*

Jeep introduced the Scrambler in 1981. The Haddam Neck, Connecticut, Fire Company outfitted this 1983 Jeep Scrambler with a 250-gpm pump. This truck is strange because it doesn't have a tank. *Photo by Mark A. Redman.*

This is a close-up of the 1983 FMC Fire Trac vehicle (see page 110). It has a 500-gpm pump, a 1000-gallon tank, and 200 gallons of foam. Note the Stang gun behind the booster reel. *Photo by Mark V. Carr.*

This 1983 International 4x4 with an American body is in service with the Cheyenne County Rural Fire Department in St. Francis, Kansas. Brush Engine 2 has a 225-gpm pump and a 1000-gallon tank. *Photo by Dennis J. Maag.*

Dover, Minnesota, Fire Department runs this 1983 Chev/Morysville utility body. This truck has a 200-gpm pump and a 265-gallon tank. *Photo by Frank Wegloski.*

Here, the New York City Fire Department assists the Jersey City, New Jersey, Fire Department at a large brush/rubbish fire. Satellite 3, which was one of three 1983 American LaFrance/Saulsbury hose wagons, is seen operating at this fire. *Photo by Ron Jeffers.*

This ex military 1984 Ford F-700 4x4/FTI (Fire Trucks Inc.) is now in service with the New Jersey Forest Fire Service. This unit is fitted with brush protection, a 250-gpm pump, and a 400-gallon tank. *Photo by Adam Alberti.*

Mini-pumpers are really small pumpers, however, many have four-wheel drive and some are also fitted to fight brush fires as well as structural fires. A few were specifically outfitted just to fight wildfires. The Colts Neck Fire Company #2 received this 1984 Chevy/Pierce 4x4 250-gpm mini-pumper equipped with a 250-gallon tank, a front riding position, and front-mounted ground sweeps. Ground sweeps are small nozzles fitted under a wildland firefighting apparatus to extinguish fire near the truck. The front firefighting position is more of a Midwestern and Western style of firefighting. A small length of hose is fitted on the front bumper so that a firefighter can stand on the front bumper to fight field/grass fires. *Photo by John Rieth.*

The old military M-1010 ambulance body that was mounted onto this chassis was cut down by the members of the Cedar Grove, New Jersey, Volunteer Fire Department. The 1984 Chevy 4x4 chassis now supports a 120-gpm pump and a 200-gallon tank. *Photo by Adam Alberti.*

The Lake Carmel, New York, Fire Department uses this rather plain 1984 Chevy/Stahl unit as both a brush and rescue unit. It has a 350-gpm pump and a 250-gallon tank. *Photo by Frank Wegloski.*

The Memphis Equipment Company buys and sells military trucks, as well as refurbishes ex military trucks and supplies spare parts for such vehicles. This M-35 was rebuilt in 1984 by the military. Later, the Osceola County, Florida, Fire Rescue received this truck from Memphis Equipment. Note the modified windshield, which is a trademark of upgrading by the Memphis Equipment Company. The fire department shops then converted this truck into a wildland engine by fabricating the brush protection, installing a 200-gpm pump and adding a 750-gallon tank. *Photo by Mark V. Carr.*

The US Forest Service at the Angeles National Forest (California) has this 1984 International with a B&Z body. It has a 275-gpm pump, a 500-gallon tank, and a 10-gallon foam tank. Engine 35 is stationed at Green Valley. As with the CDF style of apparatus, the US Forest Service has hundreds of wildland engines that all seem to have the same or similar style bodies. *Photo by Chuck Madderom.*

The largest Federal Wildland Firefighting organization is the Bureau of Land Management. The BLM fights fires on US lands that are not in US Forests, Fish and Wildlife Land, National Parks, or Indian Land. The BLM at Hole in the Wall, California, runs this 1984 Mercedes with a Beck body. Engine 3692 has a 120-gpm pump, a 450-gallon tank, and a plow. *Photo by Garry Kadzielawski.*

Massachusetts Forest Fire Control uses this circa 1985 International chassis with a Wayne school bus body as a wildfire crew transport bus. *Photo by Dick Bartlett.*

Milford, Massachusetts, uses this circa 1985 International S 4x4/EJ Murphy Forestry unit as Brush 1. This unit is fitted with a 500-gpm pump and a 500-gallon tank. *Photo by Dick Bartlett.*

This 1985 Chevy/Reading utility body truck has fire equipment mounted by Admiral. The Cape Elizabeth, Maine, Fire Department runs this unit equipped with a 100-gpm pump and a 250-gallon water tank. *Photo by Frank Wegloski.*

Also on the Outer Banks of North Carolina is the Chicamacomico Banks Volunteer Fire Department. Brush 501 is a 1985 GMC 4x4 Grumman mini-pumper that is used for brush fires. This truck has a 300-gpm pump and a 250-gallon tank. *Photo by John Rieth.*

Florida Forestry operates this circa 1985 FMC Fire Trac unit. It was photographed by Joe Abrams.

This rather high 1985 GMC 4x4 has a Phoenix body equipped with a high-pressure pump and a 500-gallon tank. Engine 33 is operated by the US Forest Service, San Bernardino, California, National Forest's Mormon Rocks Station. *Photo by Garry Kadzielawski.*

The CDF at the Tuolumne-Calaveras Ranger Unit runs this 1985 Ford F-700 Westates wildland engine. This is Spare Engine 4495; it carries a 300-gpm pump and a 500-gallon tank. *Photo by Garry Kadzielawski.*

Operated by Minotola Fire Co., of Buena, New Jersey, is this neat 1986 GMC 1-ton 4x4. This unit carries a 750-gpm front-mount pump and a 150-gallon tank. Built by the New Jersey apparatus manufacturer TASC, this unit is not only used for wildfires, but can be used as a source pumper for water supply. *Photo by Scott Mattson.*

Many fire companies/departments in mostly central and southern New Jersey have outfitted their brush trucks with brush protection called the cage. These bars protect the truck from branches, limbs and trees that are pushed over to reach a fire. Most of central/southern New Jersey is relatively flat and has sandy soil. The soft pine trees that grow in the region allow these trucks to push their way into the woods to reach the fire. This 1986 GMC 4x4 1-ton chassis had a utility body placed in it and was fitted with fire equipment by JC More. JC More is a small Pennsylvanian fire apparatus builder. This unit is one of two for the Dover Township, New Jersey, Fire District #1, and has a 150-gpm pump and a 250-gallon tank. This photo shows the unit operated by East Dover Vol. Fire Company; the other unit runs out of Toms River Fire Co. #2. *Photo by John Rowe.*

Fresh from the military, this M-1008 Chevy pickup is now operated by the Glen Hope, Pennsylvania, Fire Department. This 1986 Chevy has a small slip-on pump and tank. *Photo by Patrick Shoop.*

Terryville, New York, Engine Company #1 runs this 1986 GMC flatbed. It is fitted with a Wajax-Pacific 100-gpm slip-on unit. The unit carries 250 gallons of water. *Photo by Frank Wegloski.*

Not a lot of European fire Apparatus is in service in the US, but probably the most popular is the Mercedes Unimog 4x4 chassis. This tough truck can travel in extreme conditions. This 1986 model is operated by the Jackson Mills Vol. Fire Co. of Jackson, New Jersey. The 350-gpm pump, 500-gallon tank and the brush protect was built by a small New Jersey company called Custom Steel Fabricators, who specialized in outfitting brush trucks. *Photo Dennis C. Sharpe.*

This very different 1986 Ottawa Brimont 4x4 was refurbished in 1998 by the East Prospect, Pennsylvania, Volunteer Fire Company. Attack 42 has a 750-gpm pump and a 500-gallon water tank. Also carried on this truck is 40 gallons of foam. Note the firefighting riding position behind the cab and the rear wheels turned the opposite direction of the front wheels. The pump is a rear-mounted unit. *Photo by Howard Meile.*

Liberty Road, Maryland, Volunteer Fire Department uses this little 1986 Ford Ranger 4x4 with a 250-gpm pump and a 60-gallon tank. *Photo by Frank Wegloski.*

Typical of many fire departments in and near the more eastern counties of Maryland, this 1986 Chevy 4x4 is in service with the Odenton, Maryland, Volunteer Fire Department. It has a 250-gpm pump and a 250-gallon tank. Note the slip-on pump and tank fabricated by the fire department members. *Photo by Frank Wegloski.*

We are back in Rawlins County, Kansas. This time we see the Rawlins County, Fire Rescue District #2's 1986 Chevy 4x4. The members of the fire department mounted a 250-gpm pump and a 300-gallon tank to this truck. *Photo by Dennis J. Maag.*

Stefan Farage photographed this 1986 Ford C8000 4x4 with a Ward 79 body. It has a 110-gpm pump and an 800-gallon tank. E-318 is in service with the Santa Barbara County, California, Fire Department.

Many fire departments try to have a number of uses for their apparatus. The Fairfax County, Virginia, Fire Rescue runs this 1986 Ford F-250 with several slip-on units. One slip-on unit is a pump and tank for wildland fires, the other slip-on unit carries marine firefighting and rescue gear in a box. This photo shows Brush 20 with the marine fire/rescue box pulling Boat 20, a 1986 Boston Whaler. Note the hoops for the protection of the wheel wells during brush fire operations. In addition to marine and wildfire use, the Fairfax County Fire Rescue uses this unit as a snowplow. *Photo by Mike Sanders.*

Carson City, Nevada, Fire Department operates two of these 1986 Brimont 4x4 brush units. These French-built units carry 250-gpm pumps, 250-gallon water tanks, and 20-gallon class A foam tanks. *Photo by Shaun P. Ryan.*

This 1986 Cessna 206 is used by the New Jersey Forest Fire Service. It is used for observation and detection. *Photo by John Rieth, NJ Forest Fire Service.*

This 1987 GMC pickup for Franklin Township, New Jersey, has a small slip-on unit with a 150-gpm pump and a 250-gallon tank. Also, some brush protection was added to the front. The Iron hoops that run around the wheel wells prevent branches or limbs from getting caught in the wells. As with many wildland firefighting units in the East, this unit is equipped with a winch. *Photo by Scott Mattson.*

Foxborough, Massachusetts, is the home of the professional football team, the New England Patriots. Also home to Foxborough is this 1987 GMC 4x4 with an American Eagle body. This unit has a 750-gpm pump and a 500-gallon tank. The body has a good deal of diamond plating to protect the truck from damage from branches and brush. *Photo by Robert W. Fitz Jr.*

Pocono Summit, Pennsylvania, Fire Department runs this ex military 1987 Hummvee. It has a 150-gallon plastic tank and a small pump. *Photo by Patrick Shoop.*

Baltimore County, Maryland, has a large career and volunteer fire department. Here we see Brush 18, operated by the career firefighters in the Arndallstown section of the county. This unit is a 1987 Chevy S10 4x4 equipped with a Darley slip-on unit. The slip-on unit has a 300-gpm pump and a 75-gallon tank. Also note the snowplow attachment on the front. *Photo by Frank Wegloski.*

This 1987 International is believed to have a Warren Forestry body. This truck is operated by the Florida Department of Agriculture, Division of Forestry. It carries an early 1980s John Deere 350 dozer that pulls a trailer-mount fire line plow. *Photo by Mark W. Carr.*

Another Nanjemoy Volunteer Fire Department brush unit is this 1987 Ford F-600 4x4 with a 3-D body. It has a 300-gpm pump, a 300-gallon water tank, and a 30-gallon tank of foam. *Photo by Mike Defina Jr.*

At the Alligator River National Wildlife Refuge, the North Carolina Forestry Department is contracted by the Federal Government to provide wildfire suppression to both the wildlife refuge and the Air Force and Navy bombing ranges. This 1987 International is owned by the Air Force, but the 2000-gallon tanker was built by the shops of the North Carolina Forestry Dept. This unit has a 150-gpm pump and is operated by North Carolina Forestry at the Stumpy's Point section of the Alligator River Wildlife Refugee in Dare County. *Photo by John Rieth.*

Grumman Emergency Products built eight of these 1987 International Grumman wildland engines for Ventura County, California, Fire Department. Here we see Engine 343, which has a 500-gpm pump and a 500-gallon tank. *Photo by Chuck Madderom.*

CDF runs a number of Ford C chassis fire apparatus. This 1987 Ford C has a 1989 G. Paoletti body, a 50-gpm pump, a 650-gallon tank, and a 10-gallon tank of class A foam. Engine 1483 is from the Hot Springs Road Station in Cloverdale, California. *Photo by Shaun P. Ryan.*

Chuck Madderom captured this 1987 Ford F-700 with a 1989 Phoenix body on film. Engine 50 is operated by the US Forest Service from the Cabazon Station in the San Bernardino National Forest. It has a 325-gpm pump and a 500-gallon tank. It is important to keep in mind that some of the National Forests are as large as some states in the East! Many people who have never visited the West don't realize just how big some of America's natural lands are.

Especially in the West, large wildfires require hundreds, if not thousands, of firefighters. Sometimes the use of hand crews of 20 firefighters per crew are the only way to fight fire in hard to reach or remote areas. To transport hand crews and their equipment, many fire agencies in the West have these crew transportation trucks. More like buses than trucks, the vehicles have a very important job at wildfires. The Orange County, California, Fire Department crew 57 in Laguna Niguel, California, uses this 1987 GMC with a Master body. *Photo by Chuck Madderom.*

The Port Jefferson, New York, Fire Department has this 1988 Chevy 4x4 equipped with a body from Eastern Welding Truck Bodies. This unit has a 75-gpm pump, a 250-gallon tank, and of course the painted shark's teeth on the front bumper. *Photo by Ron Jeffers.*

The Philadelphia, Pennsylvania, Fire Department has two grass fire units in service. This one is a 1988 Chevy 4x4 with fire equipment by Saulsbury. Grass Fire Unit 3 has a 200-gpm pump and a 250-gallon tank. The city has thousands of acres of marsh grass along the banks of the Delaware River. *Photo by Jack Wright.*

This type of wildland engine is very popular with fire departments in Florida. This 1988 International 4x4 has a Southern Coach body with a 500-gpm pump and a 500-gallon tank. Note the step behind the cab leading to the firefighting riding position. Woods 31 is one of many wildland engines operated by the Orange County, Florida, Fire Department. *Photo by Glenn B. Vincent.*

This mean-looking Mack just may scare the fire out! Evergreen, Colorado, Fire Department's Tender 75 is a 1988 Mack RM with a locally built body. It has a 750-gpm pump and a 1500-gallon tank. *Photo by Rick Davis.*

This 1989 GMC 4x4 pickup was fitted with a brush cage, a 300-gpm pump and a 300-gallon tank. PL Custom, a local ambulance/rescue truck builder, built this unit for the Laurelton Fire Co. of Brick, New Jersey. Later, PL Custom would build a full-sized pumper for Laurelton. *Photo by John Rowe.*

This 1989 AM General M-931 truck tractor was converted into a tanker/water tender in 1999 by the New Jersey Forest Fire Service. This truck is a variant of the M-939 cargo truck. The series included more than 13 variants. All were produced from 1985-1991. Some trucks in this series were also manufactured by a government contractor called BMY, of York, Pennsylvania. On this rig the tanker body has been passed down from two older chassis. A 300-gpm pump and a 1200-gallon tank provide the firefighting capabilities. *Photo by John Rieth, NJ Forest Fire Service.*

The Plymouth, Massachusetts, Fire Department operates several brush-breakers. To support these breakers, this 1989 Mack DM 2500-gallon tanker with a Saulsbury body is in service as Tanker 1. *Photo by Robert W. Fitz Jr.*

This unusual brush-pumper is a 1989 International 4x4/Pierce 400-gpm unit. It carries 600 gallons of water and is in service with the Coram, New York, Fire Department. *Photo by Frank Wegloski.*

Jeep CJs are not as popular for wildland firefighting as they were. This is one of the newest ever seen by the author. This 1989 Jeep Wrangler 4x4 with a 50-gpm pump and a 150-gallon tank is in service with the Fairdale, Kentucky, Fire Department. *Photo by Greg Stapleton.*

Orange County, California, Dozer Transport 1 has in service this 1989 Western Star tractor hauling a 1989 Cat dozer. *Photo by Chuck Madderom.*

Los Angeles County operates a number of brush fire patrol units. Here, patrol 92 uses this 1989 GMC with a Royal body. It has a 125-gpm pump and a 150-gallon tank. *Photo by Chuck Madderom.*

This 1989 GMC 4x4 of the Whitney, Idaho, Fire Department has a 125-gpm pump and a 200-gallon tank. This unit also serves as a rescue squad. *Photo by Deran Watt.*

This 1989 Cat D5 dozer has a trailer-mounted fire plow. The blade is removed and only placed on the dozer when needed. It is operated by the North Carolina Forestry at Stumpy's Point. *Photo by John Rieth.*

The National Park Service operates a very large fleet of fire apparatus. Most are wildland engines, but some parks also have structural apparatus. Here, at the Joshua Tree National Monument's Rock Camp Station (California), the National Park Service runs this 1989 International with a 1990 Boise Mobile Equipment body. Engine 3632 has a CAFS unit, a 250-gpm pump and a 500-gallon tank. *Photo by Chuck Madderom.*

This 1990 Chevy 2500 pickup has a 15-gpm high-pressure pump and a 100-gallon tank. It is used by the Kentwood, Michigan, Fire Department. *Photo by Daniel A. Jasina.*

Hundreds of similar units have been built like this truck for both municipal fire companies/departments and the New Jersey Forest Fire Service. This 1990 Dodge 350 one-ton 4x4 is fitted with a 250-gpm pump and a 250-gallon tank. The New Jersey Forest Fire Service operated more than 75 such units at this time. Each truck was outfitted by New Jersey Forest Fire Service's own shops. A utility body is installed by the shops, the brush cage, tank, and other equipment is either manufactured or installed by shop forces. *Photo by John Rieth, NJ Forest Fire Service.*

Garry Kadzielawski photographed this ex Serra Pacific Power Company Ford C 4x4 chassis. In 1990 the fire department members converted this truck into a wildland engine equipped with a 350-gpm pump and a 750-gallon tank.

The US Forest Service's Engine 52, from the Sequoia National Forest (California), is a 1990 Ford F-800 with a Boise Mobile Equipment body. This unit has a 90-gpm pump and a 300-gallon tank. *Photo by Chuck Madderom.*

Rialto, California, Fire Department runs this 1990 4-door 4x4 International with a Central States body as Tender 204. This unusual truck has a 500-gpm pump and a 1500-gallon tank. *Photo by Chuck Madderom.*

This circa 1990 International roll-back, operated by the Maryland Department of Natural Resources, carries a circa 1990 Bombardier tracked vehicle fitted with a front blade and a fire line plow. *Photo by John D. Floyd.*

Both the Los Angeles City and County run Dozer Teams (units). Here, Los Angeles County Fire Department's Dozer Team 1 is a 1990 Volvo/GMC/White hauling a large 1968 Cat D8H dozer. In the late eighties, the Volvo/White Truck Company bought the rights and the name of GMC Heavy Trucks, known then as the Truck and Bus Division of GM. For the first few years, the name was changed from Volvo/White to GMC/Volvo. Later, in the 1990s, Volvo dropped both the White and GMC name brands from their truck emblems. By 2001 Volvo bought the Renault Company, who also owns Mack Trucks. *Photo by Chuck Madderom.*

The US Bureau of Land Management, at Conway Summit, California, uses this 1990 International with a 1991 Boise Mobile Equipment Company body. Engine 3132 carries a 300-gpm pump, a 500-gallon tank, and a 10-gallon foam tank. *Photo by Garry Kadzielawski.*

Another US Forest Service wildland engine from the Angeles National Forest is this 1990 International with a 1991 Boise Mobile Equipment body. Engine 25 is stationed at the Lower San Antonio Station. It has a 375-gpm pump, a 500-gallon tank, and a 15-gallon foam tank. *Photo by Chuck Madderom.*

The Marco fire apparatus company has only a few units in service in New Jersey. One of their first in New Jersey was this 1991 Ford F-350 equipped with a brush cage, a high pressure pump and a 200-gallon tank. Note the small yellow high-pressure hose on the reels. This unit is in service with the Derby Fire Co. of Bordentown Township, New Jersey. *Photo by Scott Mattson.*

This 1991 Ford F-700 was built by the New Jersey Forest Fire Service. It has a 250-gpm pump and a 1200-gallon tank. *Photo by John Rieth, NJ Forest Fire Service.*

The Monmouth County, New Jersey, Reclamation Center Fire Brigade operates this 1968 Diamond T M-54 6x6 that was refurbished in 1991 by Walker. Walker fitted a new 3000-gallon elliptical tanker body equipped with a 200-gallon foam tank and remote turret. The reclamation center is a recycling center and landfill that has acres of forest around the landfill. This unit provides fire protection to the structures, landfill and forest areas. In addition, the unit has large quick-dumps so that the unit can operate on mutual aid with the Monmouth County Tanker Taskforce. *Photo by Scott Mattson.*

This 1991 Ford F-350 4x4/Becker was reported to be the first fire apparatus in New Jersey equipped with a compressed air foam system (CAFS). The East Millstone Fire Co. of Franklin Township, New Jersey, placed this unit into service equipped with a 350-gpm pump and a 180-gallon water tank, as well as a CAFS unit. This unit is also used for structural fires. Becker Fire Apparatus used to be a dealer for E-ONE Fire Apparatus and began to manufacture their own bodies. The good eye can see the resemblance to E-ONE bodies. By 1999 Becker became affiliated with American LaFrance. *Photo by John Rieth.*

Many members of the New Jersey Forest Fire Service are also volunteer firefighters with municipal fire companies/departments, thus many municipal fire departments operate units that are built similar to the Forest Fire Service units. This 1991 Dodge 350 4x4 has a utility body, fire equipment and cage built by a local company called S&S. The owner of S&S was previously employed with the New Jersey Forest Fire Service. S&S would build brush trucks for many municipal fire departments. Later, the name and the company changed to G&S. This unit is in service with the Cecil Fire Co. of Monroe Township, New Jersey. *Photo by John Rieth.*

This 1991 Ford F-350 4x4 small-sized brush-breaker was built by the EJ Murphy Co., who has built a large number of breakers over the years. This unit is in service with the Plympton, Massachusetts, Fire Department. It has a 250-gpm pump and a 300-gallon tank. *Photo by Dick Bartlett.*

The US Army received several of these 1991 International 6x6 brush tankers. The bodies were built by KME, and were equipped with 250-gpm pumps and 1250-gallon tanks. This unit was stationed at the Fire Department at the US Military Academy at West Point, New York. *Photo by Stafan Farage.*

This gray colored 1991 Chevy 4x4 is in service with the Masontown, Pennsylvania, Fire Department. This rig has a Reading utility body, a 150-gpm pump and a 300-gallon tank. *Photo by David Bowen.*

Delray Beach, Florida, Fire Rescue runs this 1991 International Southern Coach 500-gpm brush unit. It carries 600 gallons of water. Southern Coach is now part of American LaFrance Companies. Note the angle of departure at the rear of the truck. *Photo by Mark V. Carr.*

This 1991 GMC 4x4 has a 4-Guys body. It serves the Prince Frederick, Maryland, Volunteer Fire Department. It has a 250-gpm pump and a 200-gallon tank. *Photo by Frank Wegloski.*

In the 1980s, Southern Coach, now part of the American LaFrance Companies, produced the Rhino brush units. This 1991 International 4x4 Southern Coach is in service with the Palm Beach County, Florida, Fire Department. Brush 3213 has a 500-gpm pump and a 750-gallon tank. American LaFrance now calls these units the Brush Master Series. *Photo by Mark V. Carr.*

They say don't judge a book by its cover. The California Department of Forestry (CDF) runs this experimental Model 9 engine. This unit is a 1991 Spartan with a Westates body. It is a wildland engine. Yes, it looks like a class A structural engine, but this unit has a 500-gpm Darley hydrostatic pump (to give the truck the ability to pump and roll) and a 750-gpm pump. The CDF has four of these units. *Photo by Chuck Madderom.*

The US Bureau of Land Management Dozer Team 3182 has this 1991 Ford LT9000 tractor hauling a 1990 CAT D6H high tracked dozer. Most modern Cat dozers have these high tracks. This unit is based at the BLM's Kernville Station in California. *Photo by Chuck Madderom.*

The Red, White & Blue Fire Department of Breckenridge, Colorado, has this 1991 GMC 4x4 chassis with fire equipment built by fire department members. It has a 60-gpm pump and a 200-gallon tank. Too bad it is not painted red, white, and blue! *Photo by Dennis J. Maag.*

Fort Worth, Texas, Fire Department runs this 1991 International 4x4 E-ONE as Brush 25. It has a 1000-gpm pump and a 500-gallon tank. Note the front bumper turret. *Photo by Mark V. Carr.*

Middletown Township, New Jersey, Fire Department is said to be the world's largest all volunteer municipal (not county) fire department. There are 11 companies that run from 12 stations, which roster over 600 volunteers who operate 22 pumpers, 3 ladders, 7 brush units, 2 mini-pumpers and over 20 support, command, or utility units. This 1992 Ford F-350 4x4 is one of seven brush units. This unit is operated by the Navesink H&L Co. Brush 112 is fitted with a skid-mounted 250-gpm pump and a 250-gallon tank mounted to a flat bed. The unit was put together by American Rural Fire Apparatus. In 1999 the unit received heavy damage in a wreck, the body was removed and a new utility style body was mounted. A new 200-gallon tank was fitted as well as water rescue equipment. *Photo by John Rieth.*

This Peterbuilt S&S tender was built for North Tree Fire Services in 1991. This unit has a 1000-gpm pump and a 3800-gallon tank. *Photo by S&S Fire Apparatus Company.*

In America we believe in freedom—freedom to design and use fire apparatus the way we want to. Thus, the Melrose Hose Co. of Sayreville, New Jersey, operates this 1992 International/Pierce 4x4 pumper. The unit has a 1250-gpm pump and a 500-gallon tank. The unit is also used for wildfires with the use of the front riding position on the front bumper. *Photo by John Rieth.*

The Stoughton, Massachusetts, Fire Department runs this 1992 Ford 4x4 with a C&S body as Brush Truck 1. This unit has a 350-gpm pump and a 400-gallon tank. *Photo by Robert W. Fitz Jr.*

Massachusetts State Forest Fire Control operates this 1992 International 4x4/ EJ Murphy brush-breaker with a 600-gpm pump and a 700-gallon tank. This unit is stationed at West Tisbury, on Martha's Vineyard, Massachusetts. *Photo by Stefan Farage.*

Greenwood, Delaware, runs this green and white 1992 4-door 4x4 Chevy pickup. It has a slip-on pump and tank. The pump is rated at 200-gpm and the tank carries 250 gallons of water. *Photo by Frank Wegloski.*

At Walt Disney World, Florida, the Fire and Rescue service, as well as public works, is provided by a quasi-public form of government known as the Reedy Creek Improvement Authority. Woods 31 is a 1992 Ford E-ONE 250-gpm brush unit operated by the Reedy Creek Emergency Services. *Photo by Mark V. Carr.*

Off-roading 4x4 fans should enjoy the large mud tires and lift kit installed on this 1992 Ford F-350 with a Monroe body. This Homer Township Fire Protection District, Illinois, Brush Unit 8 has a 150-gpm pump and a 250-gallon tank. *Photo by Mike Tenerelli.*

Orange County, California, runs this 1992 Ford F-350 4x4 with a Fire Bann body as Patrol 18. It has a 50-gpm pump and a 150-gallon tank. *Photo by Chuck Madderom.*

The California Office Emergency Services provide fire apparatus to county and municipal fire departments throughout the state. These units are used to assist other fire departments during large-scale emergencies, like a wildfire or an earthquake. The fire department that the engine is assigned to can use them as regular rigs, or just as Mutual Aid rigs. Here is the Napa, California, Fire Department's OES engine. It is a 1992 Spartan Westates 1000-gpm pumper fitted with a 750-gallon tank and an 8-gallon class A foam tank. *Photo by Shaun P. Ryan.*

Orange County, California, Fire Department runs this 1992 Ford F-800 Smeal wildland engine. It has a 500-gpm pump and a 500-gallon tank. *Photo by Chuck Madderom.*

As in the East, the West has a number of regional apparatus makers. Here we see the Santa Barbara County, California, Fire Department's 1992 International 4x4 West Mark 500-gpm wildland engine. It has a 500-gallon tank and is stationed in Goleta, California. *Photo by Stefan Farage.*

Forestry 1 of the Casco, Maine, Fire Department is a 1992 GMC flatbed with a 250-gpm pump and a 250-gallon tank mounted into the bed. Today, fire departments across the nation are using flatbeds equipped with pumps, tanks, and sometimes toolboxes, as a more cost-effective option for wildland fire apparatus. *Photo by Frank Wegloski.*

Gateway National Park has several units located in New York City and Sandy Hook, New Jersey. Each unit operates wildland firefighting apparatus. Sandy Hook also operates two structural engines. Several Chevy/GMC 4x4s are in service and some have pickup bodies, one has a flat bed, and others have utility bodies. This unit is a 1993 Chevy 35-gpm pump with a 250-gallon slip-on unit. *Photo by John M. Calderone.*

Coram, New York, also operates this 1993 Hummer 4x4. It was outfitted with a 250-gpm pump, a 275-gallon water tank, and a 30-gallon foam tank by a local builder. It was one of the first Hummers to have a brush cage. *Photo by Robert P. Vaccaro.*

Mark A. Redman photographed this 1993 International with a Ferrara body. This unit is operated by the Florida Department of Forestry. It serves the Big Pine Key area. It has a 250-gpm pump and a 750-gallon tank.

Advanced Four Wheel Drive and Utah-LaGrange team up to build this wildland engine on a 1993 Ford Super Duty chassis. It was delivered to the Jefferson R7 Fire Protection District, of Jefferson County, Missouri. Brush Truck 7218 has a 300-gpm pump, a 300-gallon tank, and five gallons of class A foam. *Photo by Dennis J. Maag.*

Union, Missouri, has this 1993 Chevy 4x4 Region Welding wildland engine in service; it carries a 175-gpm pump and a 200-gallon tank. *Photo by Dennis J. Maag.*

This strange water tender has a rear-mounted 500-gpm pump as well as a penalty box (a.k.a. crew seating area) behind the cab. Napa County, California, Fire Department Water Tender 10 has an 1800-gallon water tank and a 60-gallon foam tank. This truck was built on a 1993 Mack MR chassis with an H&W body. *Photo by Garry Kadzielawski.*

San Bernardino County, California, Fire Department has in service this 1993 White/GMC/Volvo water tender with a Klein body. Water Tender 77 has a 500-gpm pump and a 1200-gallon tank. *Photo by Chuck Madderom.*

In 1993 the Colorado State Forest Service refurbished this 1969 Kaiser M-35 2 1/2-ton 6x6. It has a 31-gpm pump, a 1000-gallon tank, and is operated for the Colorado State Forest Service by the West Routt Fire Protection District of Hayden County, Colorado. *Photo by Dennis J. Maag.*

Also operated by the Whitney, Idaho, Fire Department is this 1993 International 4x4 with a 125-gpm pump, a 750-gallon tank, and a 25-gallon foam tank. *Photo by Deran Watt.*

This 1993 Ford F-350 4x4, equipped with a pump and tank outfitted by JC Moore, was delivered to the Franklin Township, New Jersey, Fire District #1. It was originally assigned to the Elizabeth Avenue Fire Company, then later reassigned to the Millstone Valley Volunteer Fire Company. It has a 100-gpm pump and the tank carries 200 gallons of water. Note the brush protection mainly in the front. *Photo by Ron Jeffers.*

Pennsylvania Department of Conservation & Natural Resources Bureau of Forestry runs this 1994 Chevy. It has a utility body, a 125-gpm pump and a 200-gallon tank. It is operated from Clarks Summit, Pennsylvania. *Photo by John Rieth.*

Again Evergreen, Colorado, shows us a rather interesting wildland fire apparatus. Here we see Tender 76, a 1994 International Paystar 4x4 with a Front Range body, complete with all roll-up doors and a front-mounted 750-gpm pump. This unit can hold 1800 gallons of water in its tank. *Photo by Rick Davis.*

Kissimmee, Florida, Fire Department uses this 1994 Ford F-350 with the fire equipment by Elite. It has a 250-gpm pump and a 200-gallon tank. *Photo by Stefan Farage.*

After a devastating wildland-urban interface fire, the Oakland, California, Fire Department received several new apparatus to help fight such fires. This 1994 International 4x4 interface engine has a body built by Paoletti. It has a 500-gpm pump and a 500-gallon tank. The main function of an interface engine is to protect structures that are in the path of a wildfire. The interface unit will usually carry equipment to attack a wildfire, as well as a structure fire if need be. These type of rigs were first found in the West, but now can be found throughout the nation. *Photo by Garry Kadzielawski.*

Chuck Madderom photographed this 1994 International 4x4 Boise Mobile Equipment wildland engine. It is a US Bureau of Land Management rig from the Joshua Tree National Monument, California, Rock Camp Station. It has a 4-door cab, a 250-gpm pump, and a 500-gallon tank.

The US Forest Service operates one of the largest fleets of fire apparatus in the US. This 1994 Ford with a Boise Mobile Equipment body has a 500-gpm pump and a 500-gallon tank. It also has a 40-gallon foam tank. Engine 19 is assigned to the Angeles National Forest in California. *Photo by Chuck Madderom.*

CDF at Northstar, California, operates this 1994 Mack MS with a 1995 Master body. It has a 4-door cab, a 500-gpm pump, a 750-gallon water tank, and a 20-gallon foam tank. Master Body also built a similar unit for the San Bernardino County, California, Fire Department. *Photo by Garry Kadzielawski.*

Nevada Division of Forestry in Jacks Valley runs this 1994 International 4x4 with a Paoletti body. It has a 500-gpm pump, a 500-gallon tank, and a 20-gallon foam tank. *Photo by Garry Kadzielawski.*

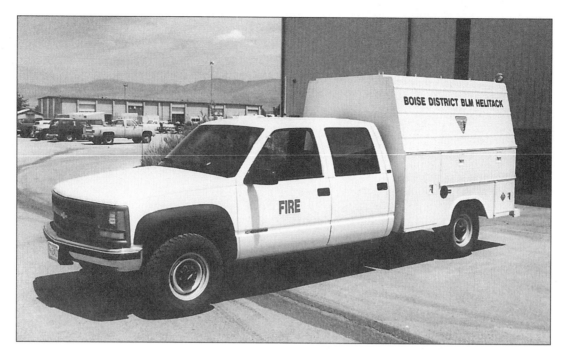

This 1994 Chevy utility truck supports the US Bureau of Land Management's Boise, Idaho, District Helitack team. This unit carries gear and equipment for the firefighting crew that operates from the helicopters. *Photo by Deran Watt.*

Six 1994 GMC chassis were built on by the New Jersey Forest Fire Service. Three of the trucks, including this one, have been turned into field repair trucks due to the poor performance from the independent suspension system, which received much damage when running over trees. Even with skid plates that are always fitted under such trucks, the low clearance of these trucks proved to be unreliable. These were the only GM trucks purchased new by the New Jersey Forest Fire Service. Some older model GM pickup chassis were obtained from the military, which have much greater ground clearance and solid axles. Due to fighting fires in the woods, the repair trucks are fitted with brush cages to cut through the woods to reach broken-down apparatus. *Photo by John Rieth, NJ Forest Fire Service.*

Another outfitter of stump-jumpers is Firematic Supply Company. This unusual 1994 International 4-door 4x4 stump-jumper was delivered to the Middle Island, New York, Fire Department. Is has since been sold. It was delivered with a CAFS unit, a 750-gpm pump, a 500-gallon water tank, and a 150-gallon foam tank. *Photo by Frank Wegloski.*

This 1995 Kenworth 6x6 was also built for North Tree by S&S Fire Apparatus. It has a 1000-gpm pump, a 3800-gallon tank, and a 200-cfm compressed air foam system. *Photo by S&S Fire Apparatus Company.*

The Howell Township Fire Company #1, otherwise known as the Adelphia Fire Company, had this 1995 Ford F-450 built not just as a brush truck but also as a mini-pumper. This unit has a 250-gpm pump, a 300-gallon water tank and carries five gallons of both class A and B foams in tanks. This company protects several trailer parks and farms that have small driveways that are too small for full-size pumpers to travel on. This unit has a hose bed on top of the booster tank in which several hundred feet of 2 1/2-inch hose is stored. In addition SCBA and other equipment for structural fires is also carried. The unit was outfitted by G&S of Tuckerton, New Jersey. *Photo by John Rieth.*

This 1995 Ford pickup has a slip-on, or skid-mounted, 100-gpm pump and a 250-gallon tank. This unit is in service with the Cheshire, Connecticut, Fire Department. Slip-on or skid-mounted pumps and tanks can be removed very easily from the truck they are mounted on! *Photo by Glenn B. Vincent.*

Bald Eagle, Pennsylvania, Fire Department has this 1995 Dodge pickup in service. Fire equipment was provided by Win-son. It has a slip-on pump and tank. *Photo by Patrick Shoop.*

The US Army at Fort Indiantown Gap, Pennsylvania, operates this 1995 International/KME 6x6 brush unit. Brush 75 has a 400-gpm pump, 1250-gallon tank, 75 gallons of class A foam, a CAFS unit and remotely controlled turret on the front bumper. *Photo by Howard Meile.*

Florida Department of Agriculture's Division of Forestry has in service this 1995 Hummvee with a 250-gpm pump and a 300-gallon tank. *Photo by Mark V. Carr.*

Michigan Department of Natural Resources operates a large fleet of wildland firefighting apparatus. This circa 1995 International 4x4 has a shop-built body and a brush cage. It is believed to have a small pump and a 500-gallon tank. *Photo by Daniel A. Jasina.*

General Safety built the body on this 1995 International chassis for the Riverside, California, Fire Department. It has a 750-gpm pump and a 500-gallon tank. *Photo by Chuck Madderom.*

Interface engines don't differ very much from wildland engines. The differences are the interface engine will usually have a larger pump and more compartment space to carry both wildland and structural fire equipment. Here is Corona, California, Fire Department's Brush 3. It seems to be more of an interface engine than a wildland engine. It has a 1000-gpm pump and a 750-gallon tank. In addition, it has a 185-gpm auxiliary pump and a body more like a structural engine. It was built in 1995 by Boise Mobile Equipment on a 1995 International 4x4 chassis. *Photo by Chuck Madderom.*

This rig is jointly operated by both the US Forest Service and the US BLM. This 1995 GMC 4x4 chassis with a Pierce truck body has a 175-gpm pump, a 200-gallon water tank, and a 5-gallon foam tank. It is part of the Upper Colorado River Interagency Fire Management in Garfield County, Colorado. *Photo by Dennis J. Maag.*

Mr. Dennis Sharpe photographed this 1996 Ford F-350 4-door cab in its environment along the bay front marsh in Avalon, New Jersey. The Avalon Volunteer Fire Department operates this unit that has an 18-horsepower high-pressure pump and a 200-gallon tank. Note the lack of brush protection on this unit. This fire department protects large tidal marsh areas, and not woods, thus the brush protection is not as necessary.

New Jersey Forest Fire Service received several 1996 Dodge chassis to build brush trucks. These were the newest style Dodges that were still in production by the spring of 2001. These were the last Dodge trucks bought new for fire suppression, ending a 40-year love affair with Dodge and the New Jersey Forest Fire Service. This 1996 model has a 250-gpm pump and a 245-gallon tank. *Photo by John Rieth, NJ Forest Fire Service.*

Pennsauken, New Jersey, a suburb community of Philadelphia, has acres of marsh along the Delaware River. To fight fires along the river this 1996 International 4x4 chassis was bought. It replaced an older chassis that this 1967 John Bean body was mounted on. Micro Fire Apparatus did the work in 1996. This truck carries a high-pressure pump and 350 gallons of water. *Photo by Ron Jeffers.*

Lacey Township, New Jersey, has thousands of acres of pine barrens in the township. The Lanoka Harbor Volunteer Fire Company operates two brush units. They are both very similar except for the 4-door cab on this 1996 Ford-F-350 4x4. The utility body was manufactured by Omaha, but the fire equipment and brush cage was manufactured by G&S who has produced more than 30 brush trucks. This unit carries a 250-gpm pump and a 300-gallon tank. *Photo by Scott Mattson.*

Several US Military bases in central/southern New Jersey operate very closely with the New Jersey Forest Fire Service. Thus, the wildland firefighting units on these bases resemble the New Jersey Forest Fire Services. This unit is one of two 1996 Dodge 4x4s equipped with a 250-gpm pump and a 200-gallon water tank, which is operated by Fort Dix Fire & Emergency Services. *Photo by Scott Mattson.*

Hummers were manufactured in a civilian version starting in 1992, and many fire departments began to order such vehicles. The Rio Grande Volunteer Fire Company of Middle Township, New Jersey, received this 1996 Hummer equipped by Fouts Brothers, a small fire apparatus manufacturer from Georgia. This unit has a 135-gpm pump and a 250-gallon water tank. *Photo by Ron Jeffers.*

The Williamstown Fire Company in Monroe Township, New Jersey, had a local welder install heavy brush protection on this 1996 Hummer chassis. This unit has a 250-gpm pump and a 300-gallon tank. *Photo By John Rieth.*

EJ Murphy has built the firefighting equipment for this 1996 Hummer. Brush 1 of the Springfield, Massachusetts, Fire Department, carries a 250-gpm pump and a 250-gallon tank. *Photo by Glenn B. Vincent.*

The Hatteras National Seashore is located along the Outer Banks of North Carolina. The National Park Service operates several wildland engines here. This 1996 Chevy 4x4 has a utility body, a 150-gpm pump and a 250-gallon tank. This unit runs out of the Nags Head section of the Hatteras National Seashore. *Photo by John Rieth.*

The Golden Gate, Florida, Fire Rescue has this 1996 Freightliner 4x4 with a Southern Coach body. Brush 71 has a 500-gpm pump, a 500-gallon tank, and a 40-gallon foam tank. *Photo by Stefan Farage.*

This circa 1996 International with a Warren Forestry body has a Dresser dozer fitted with a fire line plow. It is in service with the Georgia Forestry Commission. This photo was taken during the large wildfires in Florida in 1999. *Photo by Robert Allen.*

This 1996 Hummer has a KME body. The Dumfries-Triangle, Virginia, Fire Department runs Brush 17 at the Montclair Station. This unit has a 300-gpm pump and a 250-gallon tank. *Photo by Mike Sanders.*

The North Carolina Forestry Department operates this US Air Force 1996 Ford L9000 tractor, with a 1989 Cat D5 dozer with a trailer-mounted fire line plow. The dozer blade was not attached at this time; it was left at the forestry station and only attached on the machine when needed. These units are at the Alligator River National Wildlife Refuge, at Stumpy's Point, North Carolina. *Photo by John Rieth.*

The North Carolina Forestry Department runs several of these US Air Force 1996 Ford L9000 tractors. This unit hauls a 1983 Flex Track swamp machine. The machine carries a 50-gpm pump and a 250-gallon tank; it also pulls a trailer-mounted fire line plow. These units are at the Alligator River National Wildlife Refugee at Stumpy's Point, North Carolina. *Photo by John Rieth.*

This sharp-looking 1996 Ford chassis was retrofitted with a 1973 John Bean high-pressure pump. The tank holds 200 gallons. Brush 8739 is operated by the Cottleville, Missouri, Fire Protection District. *Photo by Dennis J. Maag.*

The National Park Service at Indiana Dunes National Lake Shore, in Chesterton, Indiana, has this 1996 Chevy 3500 4x4 with a Becker body. This unit has a 120-gpm pump, a 300-gallon tank, and a 30-gallon foam tank. *Photo by Garry Kadzielawski.*

Fire Attacker outfitted this 1996 Hummer for the Antonia, Missouri, Fire Department. It has a portable pump and a 249-gallon tank. *Photo by Dennis J. Maag.*

As we said before, the California Department of Forestry provides structural/municipal fire suppression under contract to counties and municipalities and unincorporated areas throughout the state. Here is CDF Engine 3592, a 1996 HME West Mark pumper. It has a 1000-gpm pump, a 600-gallon water tank, and a 30-gallon foam tank. The main focus of this rig is structural firefighting, however, all of the CDF units in some way are able to either fight a wildfire or provide support during a wildfire. This rig is stationed at Yucca Valley, California. *Photo by Chuck Madderom.*

In different parts of the US, various agencies are responsible for wildland firefighting. In Larimer County, Colorado, the County Sheriff's Office provides wildfire suppression and runs this 1996 International 4x4 with a Frontier body. It has a 300-gpm pump, a 700-gallon tank, and a 20-gallon class A foam tank. *Photo by Rick Davis.*

The US BLM runs this 1996 International 4x4 with a 100-gpm pump and a 700-gallon tank from the Lower Snake River District in Boise, Idaho. *Photo by Deran Watt.*

This 1996 Air Tractor 802AF aircraft is operated by a contractor for the US Forest Service. It flies from the Fort-Collins-Loveland Airport in Colorado. The aircraft can hold up to 800 gallons of water or retardant. *Photo by Rick Davis.*

This 1996 Ford Louisville chassis has a 1997 Elliot body. This US Forest Service water tender has a 450-gpm pump, a 1500-gallon water tank, and a 45-gallon foam tank. Water Tender 3 is assigned to the San Bernardino National Forest's Mill Creek Station. *Photo by Chuck Madderom.*

Due to many off-road vehicles having crashes deep in the woods, many fire companies have equipped brush trucks with extrication equipment like the jaws of life. The Cassville Fire Company of Jackson, New Jersey, received this 1997 Ford F-450. It carries a 250-gpm pump, a 300-gallon water tank, full set of extrication equipment, as well as brush protection. A local welder fabricated the brush protection. *Photo by John Rieth.*

In 1996 Oshkosh Truck Corporation and Pierce Fire Apparatus teamed up to produce the Phoenix wildland fire apparatus. The apparatus is based on the Oshkosh military Heavy Truck Expanded Mobility Tactical Truck (HEMTT) chassis. The US military calls this chassis the M-977 series, with a number of variants. This chassis had been used in the past as a base for several civilian airport crash units in the US, however, in 1996 Oshkosh and Pierce teamed up to produce a new generation of these vehicles for civilian fire agencies. This demo unit was set up for wildland firefighting, but others could be set up for ARFF (Airport Crash), or re-supply units. This demo has a 500-gpm pump, a 2500-gallon water tank, and a 50-gallon foam tank. Other firefighting options are available, including a Compressed Air Foam System (CAFS), larger pumps, under-truck nozzles, pre-connected hand lines, a Global Positioning System, Forward Looking Infrared system, in addition to other normal options like air conditioning. This unit has a top speed of 75 mph, 50-ft turning radius, can run through four feet of water, and can ascend/descend 60% grades. The truck is 70,000 pounds, and is powered by a Detroit Diesel Series 60 470-hp diesel engine. The North Trees Fire Service, which is a private contract fire suppression company, operates the Phoenix. In addition the US Army has developed the new Tactical Fire Fighting Truck (TFFT) based on this chassis. The Army's new units will have a 1000-gpm pump, a 1000-gallon water tank, and two 60-gallon tanks for both class A & B foams. *Photo and info. thanks to John Giesfeldt of Pierce Manufacturing Inc.*

This ex military M-966 series Hummer, or Humvee, was converted into a wildfire fighting unit in 1997 by the Rocky Hill, Connecticut, Fire Department. The Hummer was built starting in 1984 by the AM General Corp. until the 1990s. The series started with M-966 and 15 different variants concluded with the M-1046. Including the trailer, this unit has a 250-gpm pump and can haul 500 gallons of water. The truck is painted black. *Photo by Glenn B. Vincent.*

The Fire Department of the city of New York operates eight brush fire units throughout the city. This 1997 International 4x4/Saulsbury operated by Brush Fire Unit 1 is one of five delivered in 1997. Each unit has a 500-gpm pump and a 500-gallon tank. *Photo by John M. Calderone.*

Kimberton, Pennsylvania, Fire Company runs this pod system truck. It is a 1997 Ford F-350 4x4 chassis equipped to carry one of three pods. The pod shown is for brush fires, the other pod has rescue equipment, and the last pod is a dumpster used after a structural fire to help clean up. This unit was built by Lee's Emergency Equipment. *Photo by Dennis C. Sharpe.*

In Torrington, Connecticut, the Drakeville Volunteer Fire Department runs this very attractive 1997 Ford/Gowans-Knight 250-gpm brush unit. The truck carries a 300-gallon booster tank. *Photo by Mark A. Redman.*

S&S Fire Apparatus has built a number of water tender/wildland fire apparatus for the private contracted fire suppression company North Tree Fire Services Inc. North Tree operates throughout the West and provides training, prevention, pre-suppression, and suppression under contract. They have a large fleet of tenders, engines, and support units, as well as dozers. Here we see North Tree Unit 290, a 1997 Kenworth 6x6 chassis with an S&S body. It has a compressed air foam system, a 2600-gallon tank, two electronically controlled turrets, and a 750-gpm pump. This unit can function as a water tender equipped with a large quick dump valve in the rear and sides and a drop tank. It can also fight a wildfire and serve as an interface engine with a CASF unit. *Photo and info. by S&S Fire Apparatus Company.*

North Carolina Forestry runs a number of these Chevy pickups with slip-on pumps and tanks. This 1997 Chevy 4x4 has a 250-gpm pump and a 300-gallon tank. It is operated in the Dare County area. *Photo by John Rieth.*

This 1997 Ford F- Super Duty (a.k.a. F-450) has an EBY/Singer body and fire equipment. Brush 2 of the Purcellville, Virginia, Volunteer Fire Company has a 500-gpm pump and a 300-gallon tank. *Photo by Mike Sanders.*

The Copper Mountain, Colorado, Ski Resort has their own fire department. This 1997 AM General Hummer 4x4 has a Sutphen body. This unit has a high-pressure pump, a 70-gallon water tank, and a 10-gallon tank of class A foam. *Photo by Dennis J. Maag.*

This rather interesting fire apparatus is Rescue 12 of the Greenacres, Florida, Public Safety Department. It is a 1997 International 4x4, E-ONE 500-gpm brush pumper. This unit has a 250-gallon tank. Note the rescue pumper body complete with roll up compartment doors. This unit has a large amount of brush protection in the front of the truck. *Photo by Mark V. Carr.*

The California Department of Forestry operates Brush Engine 3163 from Riverside County, California. This unit is a 1997 International 4x4 with a West Mark body. It has a 500-gpm pump, a 500-gallon tank, and a 20-gallon foam tank. The CDF has many similar units in service throughout the state. *Photo by Chuck Madderom.*

Kern County, California, Fire Department uses this 1997 International 4x4 with a Pierce body as an interface engine. Engine 57 has a 500-gpm pump and a 500-gallon tank, and is stationed at Frazier Park, California. *Photo by Chuck Madderom.*

S&S Fire Apparatus built several of these 1997 Freightliner FL 112 tanker-pumpers for the US Bureau of Land Management. Water Tender 3192 is assigned to the Kernville, California, Station. This unit has a 500-gpm pump and a 3500-gallon tank. *Photo by Chuck Madderom.*

In Washoe County, Nevada, the Truckee Meadows Fire Protection District uses this 1997 International 4x4 Master Body Works brush unit. Brush 6 has a 500-gpm pump, a 500-gallon tank, and a 20-gallon class A foam tank. *Photo by Shaun P. Ryan.*

S&S Fire Apparatus delivered several of these 1997 Freightliner 4x4 wildland engines to the US Bureau of Land Management. BLM Engine 1946 has a 125-gpm pump, an 865-gallon water tank, and a 5-gallon foam tank. It is in service at Carlin, Nevada. S&S has used this style of apparatus geared for municipal fire departments, which are called Outland wildland engines. *Photo by Garry Kadzielawski.*

Pierce delivered this Hawk wildland engine in 1997 to Gerlach, Nevada. It has an International 4x4 chassis, 500-gpm pump, 830-gallon tank, and a 20-gallon foam tank. Note the rear compartment stops short. This makes for a better angle of departure when traveling off road. *Photo by Garry Kadzielawski.*

Pierce has built many rigs that are used to fight wildfires, but in 1997 Pierce introduced the Hawk series of wildland fire apparatus. The Hawk has a 4x4 International chassis, pumps from 300-500 gpm, tanks that hold up to 850 gallons, foam tanks and systems including a CAFS system. Two- or four-door cabs are available as well as different body styles. The truck can be set up more for wildand firefighting, or as an urban-interface unit. This 1997 International 4x4 Pierce Hawk demo has a 300-gpm bumper turret. *Photo and info. thanks to John Giesfeldt of Pierce Manufacturing Inc.*

Another fire company who had other ideas for a brush unit. The Hazlet, New Jersey, Fire Company planned this 1998 Ford F-550 as a mini-pumper. Hazlet also has several trailer parks with small streets. G&S mounted a 250-gpm pump and a 300-gallon tank. *Photo by John Rieth.*

Also in 1998, Lee's put together this Ford F-350 pickup for Waterford Township, New Jersey. This unit carries a 250-gpm pump and a 200-gallon tank, and limited brush protection. A year before, Lee's delivered a Ford-350 with a utility body and full brush protection to the Waterford Township Fire Dept. *Photo by John Rieth.*

This 1998 International 4x4 brush-breaker was being prepared for delivery to the Carver, Massachusetts, Fire Department, by the EJ Murphy Co. This unit has a 300-gpm pump and a 750-gallon tank. *Photo by Dick Bartlett.*

The Boston Fire Department runs two of these 1998 International E-ONE brush units. They each have a 500-gpm pump and a 500-gallon tank. *Photo by Dick Bartlett.*

Palm Beach County, Florida, runs a large number of these large brush units. Some are on International chassis. Others, like Brush 46, are built on Freightliner chassis. This 1998 unit has a 3D body, a 500-gpm pump, and a 750-gallon tank. Two of the largest suppliers of this type of wildland engines, 3D and Southern Coach, are now affiliated with American LaFrance. *Photo by Mark V. Carr.*

Engine 43 of the Shelton, Connecticut, Fire Department covers the Pine Rock Park area with this 1998 International 4x4 Pierce Hawk. It has a 500-gpm pump and a 500-gallon tank. *Photo by Alan Mudry.*

Lee's Emergency Equipment of Tuckerton, New Jersey, delivered two of these 1998 Ford F-450 4x4 units to Manchester Township, New Jersey. One truck went to the Manchester Township Fire Company; this one is in service with the Ridgeway Fire Company. Each unit came equipped with a 250-gpm pump and a 300-gallon tank. Note the pump mounted in the first compartment on the right side of the body, a New Jersey trademark. *Photo by John Rieth.*

Florida Division of Forestry runs this 1998 International and what is believed to be a Warren Forestry body. This truck hauls a circa 1995 Bombardier tracked vehicle. The Bombardier is set up with a firefighting package to help fight fires in swamp or marsh areas. *Photo by Mark V. Carr.*

Southern Shores, North Carolina, is on the Outer Banks. This 1998 GMC 3500 has an E-ONE American Eagle brush body, a 250-gpm pump and a 200-gallon tank. *Photo by Mike Sanders.*

Showell, Maryland, claims to have the world's largest pumper-tanker. This 1998 Volvo of North America chassis has a Southern Coach body equipped with a 1500-gpm pump and a 5000-gallon tank. Because of the weight, the body is mounted on this tri-axle chassis. This unit supplies water to any emergency—not just wildfires. *Photo by Mike Defina Jr.*

Los Angeles County Fire Department's Dozer Team 6 is this 1998 Freightliner FLD 120 tractor hauling a 1996 Cat D8N dozer. *Photo by Chuck Madderom.*

This 1998 International 4x4 chassis has a Master body. This water tender has a 300-gpm pump, a 1500-gallon tank, and a front-mounted turret on the bumper. Water Tender 301 is in service with the Hesperia, California, Fire Department. *Photo by Chuck Madderom.*

This rather short 1998 International tanker has a 1500-gallon tank and a 50-gpm pump. It has a Klein body and is in service with the Riverside County, California, Fire Department. Note the amber lens in the light bar mounted on the roof. Also, take a close look at the fire department emblem on the door—it has the California Department of Forestry emblem. Starting in the 1990s the CDF, under contract, provided the fire suppression in Riverside County. *Photo by Chuck Madderom.*

Shaun P. Ryan photographed this 1998 International 9300 tractor pulling a 1991 Load King trailer, with a 1991 Cat D-6H dozer. This is one of three in the Nevada Division of Forestry's inventory.

The Aspen, Colorado, Fire Department has this 1998 International 4x4 with a West Mark body. It has a 500-gpm pump, a 500-gallon water tank, and an 18-gallon foam tank. *Photo by Dennis J. Maag.*

The Idaho Department of Lands runs this 1998 Dodge 4x4 flatbed with a 150-gpm pump and a 200-gallon tank. *Photo by Deran Watt.*

The Tohono O'Odham Native American Reservation in Sells, Arizona, operates this 1998 Chevy 3500 HD 4x4 chassis with a Monroe body. It has a 120-gpm pump, a 200-gallon water tank, and a 5-gallon foam tank. *Photo by Garry Kadzielawski.*

Los Angeles County, California, Fire Department has this 1998 International 6x6 chassis with a 1999 Master Body. It operates as Water Tender 70 in Malibu, California. This tender has a 500-gpm pump and a 3000-gallon tank. *Photo by Chuck Madderom.*

Wayne, New Jersey, Fire Company #1 received this 1999 AM General Hummer/Fire Attacker 4x4. This truck is fitted with a 500-gpm pump, a 200-gallon water tank, and a 25-gallon tank of class A foam. In addition, a bumper turret is fitted along with a winch. During the 1990s, Fire Attacker began to market fire and rescue equipment on Hummer chassis, and have been quite successful. Today many manufactures build on Hummer chassis. *Photo by Ron Jeffers.*

Some fire departments only have a need for a pickup with a skid-mounted pump and tank. The Franklin Fire Co. of Mansfield Township, New Jersey, received such a unit built on a 1999 Ford F-350 4x4 pickup. This truck carries a 350-gpm pump and a 200-gallon tank. *Photo by Scott Mattson.*

Amtrak, the National Passenger Railroad, operates a very large train yard in Queens, New York City. To help fight brush fire in and near the yard a 1999 GMC 4x4 HI-Rail chassis, equipped with an Astoria body, was placed into service. The unit has a 250-gpm pump, a 300-gallon tank and can travel on or off road as well as on railroad tracks. *Photo by John M. Calderone.*

The US Virgin Island Fire Service has in service this 1999 International 4x4 US Tanker 500-gpm tanker-pumper. It has an 1800-gallon tank. *Photo by Mark V. Carr.*

Florida Division of Forestry has this 1999 Volvo tractor with a 500-gallon tank and a small pump, pulling a trailer with a 1999 John Deere 450 dozer hauling a trailer-mounted fire line plow. *Photo by Joe Abrams.*

This is Harrison Township, Ohio, Brush 1. It is a 1999 Chevy 4x4 4-door pickup with an American Fire slip-on unit with a 250-gpm pump and a 200-gallon tank. *Photo by Greg Stapleton.*

San Bernardino, California, Brush-Engine 228 is a 1999 International 4x4 with a 2000 Pierce body. This is a Pierce Hawk model—it has a 500-gpm pump, a 600-gallon water tank, and a 30-gallon foam tank. *Photo by Chuck Madderom.*

The CDF in Napa County, California, runs this 1999 Mack CH tractor pulling a 2000 Cozad trailer, which is hauling a 1996 Cat D5HXL dozer. *Photo by Shaun P. Ryan.*

Some things aren't what they seem! This 2000 Pierce Saber 1250-gpm pumper is not considered a pumper. The Bernardsville, New Jersey, Volunteer Fire Company uses this 4x4 chassis for both a brush unit and a rescue unit. This truck has a 430-gallon water tank and a 20-gallon foam tank, and, in addition, a full set of extrication equipment, including air bags, are loaded onto this unit. *Photo by Ron Jeffers.*

The New Jersey Forest Fire Service will be taking delivery of 29 Ford F-350s and F-450s during 2000 and 2001. This 2000 Ford F-350 has a Knaphid utility body, a 250-gpm pump, and a 250-gallon tank. The trademarked New Jersey Brush Cage and other equipment were outfitted or fabricated by New Jersey Forest Fire Shops. *Photo by John Rieth, NJ Forest Fire Service.*

The Midway Fire Department of Centreville, Illinois, uses this 2000 Ford F-450 E-ONE as a rescue. It has a 250-gpm pump, a 350-gallon tank, and 20-gallon foam tank. *Photo by Dennis J. Maag.*

Los Angeles City operates brush patrol units was well as the county. Here we see the LA City Patrol 99, which is a 2000 Ford F-350 4x4 pickup carrying a Darley 50-gpm pump, a 150-gallon water tank, a 10-gallon foam tank, and a compressed air foam system (CAFS). *Photo by Adam Alberti.*

This Forestry Crawler is a year-2000 demo unit. It is marketed by the Dig-it Company. This model D-7000 FL is equipped with a Whitfield fire plow and an 80-hp Cummings Diesel engine. *Photo by John Rieth.*

This 2000 Freightliner/KME interface engine is operated by the Jackson Mills Volunteer Fire Company of Jackson, New Jersey. It has a 1250-gpm pump, a 750-gallon tank, 60 gallons of foam, and has a 500-gpm PTO pump. It also has extrication equipment and 4x4 system. *Photo by Dennis C. Sharpe.*

This Los Angeles City Fire Department Helicopter 1 is a Bell 412 helicopter. On the bottom the aircraft has a 350-gallon water tank. The Bell 412 is a newer variant of the Bell 212, which traces its roots back to the famous Bell UH-1 Huey military helicopter. The LA City Fire Department uses six helicopters for the year 2001. *Photo by Daniel A. Jasina.*

In the spring of 2001 the shop forces of the New Jersey Forest Fire Service, converted this 1988 International 4x4/Altac cable layer (ex military) into a wildland engine. This unit has a 250-gpm pump and a 500-gallon tank. *Photo by John Rieth, NJ Forest Fire Service.*

Index

ARIZONA
Sells .... pg. 212
CALIFORNIA
Angeles National Forest ....
    pp. 125, 152, 174
Angels Camp .... pg. 111
California Department of
    Forestry .... pg. 158
Cloverdale .... pg. 140
Conway Summit .... pg. 152
Corona .... pg. 181
El Cariso .... pg. 13
Escondido .... pp. 115, 116
Goleta .... pg. 166
Griffith Park .... pg. 45
Hesperia........................ pg. 209
Hole in the Wall .... pg. 125
Joshua Tree National Monument
    pg. 174
Kern County .... pp. 45, 200
Kernville .... pp. 159, 201
Laguna Niguel .... pg. 141
Los Angeles City Fire Depart-
    ment .... pp. 218, 220
Los Angeles County Fire
    Department ....
    pp. 72, 117, 118, 146, 151,
    209, 212
Napa .... pg. 165
Napa County .... pp. 169, 216
Nevada City .... pg. 5
Northstar .... pg. 175
Oakland .... pg. 173
Ontario .... pg. 88
Orange County ....
    pp. 146, 164, 165
Perris .... pg. 116
Redwood Valley Calpella . pg. 60
Rialto .... pg. 150
Riverside .... pg. 181
Riverside County ... pp. 200, 210
Rock Camp Station pp. 148, 174
San Bernardino ..... pp. 118, 128,
    216
San Bernardino County . pg. 170
San Bernardino National Forest
    pp. 107, 140, 193
San Diego .... pg. 102
San Ramon Valley .... pg. 117
Santa Barbara County ... pg. 134
Sequoia National Forest pg. 150
Temecula....................... pg. 106
Tuolumne-Calaveras ..... pg. 129
Ukiah .... pg. 71
Ventura County .... pg. 139
Winters .... pg. 35
Yucca Valley .... pg. 191
COLORADO
Aspen............................ pg. 211
Basalt .... pg. 25
Breckenridge................. pg. 159
Copper Mountain .... pg. 199
Evergreen .... pp. 61, 143, 172
Fort-Collins-Loveland Airport pg. 192
Garfield County .... pg. 182
Hayden County .... pg. 170
Larimer County .... pg. 191
CONNECTICUT
Avon .... pg. 52
Cheshire....................... pg. 178
Colchester .... pg. 19
Connecticut Forest Fire Service
    pg. 67
Ellington .... pg. 47
Glastonbury .... pp. 64, 96
Haddam Neck .... pp. 51, 120
Long Hill .... pg. 47
Pawcatuck .... pg. 52
Quinebaug .... pg. 40
Rocky Hill .... pg. 195
Sandy Hook .... pg. 93
Shelton .... pp. 78, 206
Simsbury .... pg. 119
Torrington .... pg. 196
Westfield .... pg. 17
DELAWARE
Dover .... pg. 42
Greenwood .... pg. 163
Port Penn .... pg. 65
FLORIDA
Cecil Field .... pp. 110, 120
Delray Beach .... pg. 157
Florida Department of Agricul-
    ture .... pp. 138, 180
Florida Department of Forestry
    pp. 73, 128, 168, 207, 215
Golden Gate .... pg. 187
Greenacres .... pg. 199
Kissimmee .... pg. 173
Orange County .... pg. 142
Osceola County ..... pp. 105, 124
Palm Beach County pp. 158, 206
Walt Disney World .... pg. 163
GEORGIA
Georgia Forestry Commission
    pg. 187
HAWAII
Honolulu .... pg. 119
IDAHO
Boise .... pp. 176, 192
Idaho Department of Lands .. pg. 211
Whitney .... pp. 147, 171
ILLINOIS
Centreville .... pg. 218
Fairview Heights .... pg. 81
Homer Township .... pg. 164
Maeystown .... pg. 30
Mt. Vernon .... pg. 105
Waterloo .... pg. 92
INDIANA
Carlisle .... pg. 44
Chesterton .... pg. 190
Elmore Township .... pg. 91
Jasonville .... pg. 94
Martin County .... pg. 55
Plainville .... pg. 24
St. Joseph Township .... pg. 85
Toaktown .... pg. 99
Veale Township .... pg. 38
Washington .... pg. 43
KANSAS
Rawlins County .... pp. 49, 133
St. Francis .... pg. 121
KENTUCKY
Battletown .... pg. 102
Burgin .... pg. 76
Fairdale .... pg. 145
MAINE
Brunswick .... pg. 75
Buxton .... pg. 22
Cape Elizabeth .... pg. 127
Casco .... pg. 166
Cornish .... pg. 96
Goodwins Mills................ pg. 23
Kennebunk .... pg. 31
Kezar Falls .... pg. 75
Lebanon .... pg. 22
Maine Forest Fire Service pg. 26
Newfield .... pg. 8
Wells .... pg. 50
MARYLAND
Annapolis .... pg. 12
Anne Arundel County ...... pg. 11
Arcadia .... pg. 101
Baltimore County .... pg. 137
Glyndon.......................... pg. 79
Liberty Road .... pg. 132
Maryland Department of Natural
    Resources ... pp. 65, 80, 94, 151
Mt. Airy .... pg. 59
Nanjemoy .... pp. 43, 138
Odenton .... pg. 133
Prince Frederick ...... pp. 87, 157
Romancoke .... pg. 80
Showell .... pg. 208
Sudlersville .... pg. 14
West Annapolis .... pg. 20
MASSACHUSETTS
Barnstable County Forest Fire
    Service .... pg. 21
Bedford .... pg. 14
Boston .... pg. 205
Carver .... pp. 50, 205
Department of Environmental
    Management .... pg. 112
Foxborough .... pg. 136
Leicester .... pg. 104
Marstons Mills .... pg. 51
Mashpee .... pg. 28
Massachusetts Department of
    Forestry .... pp. 56, 62
Massachusetts Forest Fire
    Control .... pg. 126
Milford .... pg. 126
Otis Air Force Base .... pg. 93
Plainville .... pg. 66
Plymouth .... pg. 144
Plympton .... pg. 155
Springfield .... pg. 186
Stoughton .... pg. 162
West Tisbury .... pg. 162

MICHIGAN
- Comstock .... pg. 71
- Kentwood .... pg. 148
- Michigan Department of Natural Resources .... pg. 180
- Richland Township ..... pp. 36, 98

MINNESOTA
- Dover .... pg. 121

MISSOURI
- Antonia .... pg. 190
- Cottleville .... pg. 189
- Jefferson County ...... pp. 91, 168
- Mapaville .... pg. 33
- St. Clair .... pg. 110
- Union .... pg. 169
- Wentzville .... pg. 60

NEVADA
- Carlin .... pg. 202
- Carson City .... pp. 92, 135
- Gerlach .... pg. 202
- Nevada Division of Forestry . pp. 72, 86, 175, 210
- Spring Creek .... pg. 103
- Storey County .... pg. 99
- Washoe County .... pp. 73, 201

NEW HAMPSHIRE
- Hooksett .... pp. 19, 115

NEW JERSEY
- Avalon .... pg. 182
- Bayville .... pg. 9
- Berkeley Township .... pg. 32
- Bernardsville .... pg. 217
- Bordentown Township ... pg. 153
- Brick .... pp. 21, 143
- Buena .... pp. 112, 129
- Cedar Grove .... pg. 123
- Centerton .... pg. 16
- Colts Neck .... pp. 18, 123
- Delran .... pg. 39
- Deptford Township .... pg. 15
- Dover Township .... pp. 38, 130
- Downstown .... pg. 111
- Earle .... pg. 95
- Eatontown .... pg. 88
- Farmingdale .... pg. 10
- Franklin Township . pp. 136, 154, 171
- Hazlet .... pp. 114, 204
- Howell Township ...... pp. 95, 178
- Island Heights .... pg. 63
- Jackson .... pp. 131, 193, 219
- Jamesburg .... pg. 11
- Jersey City .... pg. 122
- Lacey Township ... pp. 7, 30, 184
- Little Egg Harbor .... pg. 33
- Manalapan .... pg. 100
- Manchester Township .... pg. 207
- Mansfield Township .... pg. 213
- Marlboro Township ..... pp. 27, 31
- Matawan Township .... pg. 39
- Middletown Township .. pp. 100, 160, 185
- Millstone Valley 171
- Monmouth County .... pg. 154
- Monroe Township ... pp. 155, 185
- Moonachie .... pg. 108
- Mullica Township .... pg. 74
- Neptune .... pg. 74
- New Jersey Forest Fire Service pp. 5, 6, 7, 15, 46, 49, 55, 61, 62, 66, 82, 87, 103, 107, 108, 122, 135, 144, 149, 153, 176, 183, 217, 220
- North Hudson .... pg. 82
- Oxford .... pg. 23
- Pennsauken .... pg. 183
- Pine Beach .... pg. 77
- Plainsboro .... pg. 63
- Sandy Hook .... pg. 167
- Sayreville .... pp. 113, 161
- Tinton Falls .... pg. 34
- Transboro .... pg. 89
- Waterford Township .... pp. 8, 27, 204
- Wayne .... pg. 213
- Woodland Township .... pg. 89
- Winslow Township .... pg. 77

NEW YORK
- Bedford .... pg. 40
- Bridgehampton .... pg. 68
- Bronx .... pg. 37
- Coram .... pp. 145, 167
- Deer Park .... pg. 41
- East Hampton .... pg. 57
- Farmingville .... pg. 67
- Hawthorne .... pg. 109
- Kent Cliffs .... pg. 25
- Lake Carmel .... pg. 124
- Lindenhurst .... pg. 68
- Medford .... pg. 53
- Middle Island .... pg. 177
- Montauk .... pg. 17
- New York .... pg. 195
- Port Jefferson .... pg. 141
- Quaker Springs .... pg. 12
- Queens .... pg. 214
- Ridge .... pp. 20, 36
- Staten Island .... pg. 32
- Terryville .... pg. 131
- Thornwood .... pg. 41
- West Point .... pg. 156
- Yaphank .... pg. 24

NORTH CAROLINA
- Chicamacomico Banks .. pg. 127
- North Carolina Forestry Department .... pp. 139, 198
- Outer Banks of North Carolina pp. 70, 114, 186
- Southern Shores .... pg. 208
- Stumpy's Point .... pp. 147, 188, 189

OHIO
- Harrison Township .... pg. 215
- Miami Township .... pg. 44
- Trvro Township .... pg. 29

PENNSYLVANIA
- Aultman .... pg. 57
- Bald Eagle .... pg. 179
- Barto .... pg. 37
- Bendersville .... pg. 58
- Boyertown .... pg. 69
- Brady Township .... pg. 70
- City of Reading .... pg. 109
- Clarks Summit .... pg. 172
- Coalport .... pg. 84
- Cooperstown .... pg. 78
- Earl Township .... pg. 83
- East Prospect .... pg. 132
- Forksville .... pg. 13
- Fort Indiantown Gap ... pg. 179
- Fryburg .... pg. 79
- Glen Hope .... pg. 130
- Glen Iron .... pg. 85
- Hamburg .... pg. 42
- Hamlin .... pg. 90
- Kimberton .... pg. 196
- Kreamer .... pg. 76
- Littlestown .... pg. 54
- Liverpool .... pg. 58
- Madera .... pg. 97
- Masontown .... pg. 156
- Mineral Point .... pg. 83
- Mount Penn .... pg. 104
- New Ringgold .... pg. 84
- Newville .... pg. 113
- Oil City .... pg. 48
- Penn Forest Township ..... pg. 18
- Penns Creek .... pg. 28
- Philadelphia .... pg. 142
- Pocono Summit .... pg. 137
- Suedburg .... pg. 64
- Upper Salford Township .. pg. 48
- Virginville .... pg. 69
- Waterville .... pg. 53

RHODE ISLAND
- Scituate .... pg. 34
- Nooseneck Hill .... pg. 56

TENNESSEE
- Caton's Chaple/Richardson Cove .... pg. 29
- St. James .... pg. 90

TEXAS
- Fort Worth .... pg. 160

US VIRGIN ISLANDS pg. 214

VIRGINIA
- Ashburn .... pg. 98
- Dumfries-Triangle .... pp. 97, 188
- Fairfax County .... pg. 134
- Front Royal .... pg. 54
- Purcellville .... pg. 198
- Reva .... pg. 59
- Weyers Cave .... pg. 101

WASHINGTON
- East Olympia .... pg. 81

WEST VIRGINIA
- Jefferson County .... pg. 35
- Wiley Ford .... pg. 16

WISCONSIN
- Bristol .... pg. 106

WYOMING
- Laramie County .... pg. 86

MORE TITLES FROM ICONOGRAFIX:

AMERICAN CULTURE
AMERICAN SERVICE STATIONS 1935-1943	ISBN 1-882256-27-1
COCA-COLA: A HISTORY IN PHOTOGRAPHS 1930-1969	ISBN 1-882256-46-8
COCA-COLA: ITS VEHICLES IN PHOTOGRAPHS 1930-1969	ISBN 1-882256-47-6
PHILLIPS 66 1945-1954	ISBN 1-882256-42-5

AUTOMOTIVE
CADILLAC 1948-1964	ISBN 1-882256-83-2
CAMARO 1967-2000	ISBN 1-58388-032-1
CLASSIC AMERICAN LIMOUSINES 1955-2000	ISBN 1-882256-041-0
CORVAIR by CHEVROLET EXPERIMENTAL & PRODUCTION CARS 1957-1969 LUDVIGSEN LIBRARY SERIES	ISBN 1-58388-058-5
CORVETTE THE EXOTIC EXPERIMENTAL CARS, LUDVIGSEN LIBRARY SERIES	ISBN 1-58388-017-8
CORVETTE PROTOTYPES & SHOW CARS	ISBN 1-882256-77-8
EARLY FORD V-8S 1932-1942	ISBN 1-882256-97-2
IMPERIAL 1955-1963	ISBN 1-882256-22-0
IMPERIAL 1964-1968	ISBN 1-882256-23-9
LINCOLN MOTOR CARS 1920-1942	ISBN 1-882256-57-3
LINCOLN MOTOR CARS 1946-1960	ISBN 1-882256-58-1
PACKARD MOTOR CARS 1935-1942	ISBN 1-882256-44-1
PACKARD MOTOR CARS 1946-1958	ISBN 1-882256-45-X
PONTIAC DREAM CARS, SHOW CARS & PROTOTYPES 1928-1998	ISBN 1-882256-93-X
PONTIAC FIREBIRD TRANS-AM 1969-1999	ISBN 1-882256-95-6
PONTIAC FIREBIRD 1967-2000	ISBN 1-58388-028-3
STUDEBAKER 1933-1942	ISBN 1-882256-24-7
ULTIMATE CORVETTE TRIVIA CHALLENGE	ISBN 1-58388-035-6

BUSES
BUSES OF MOTOR COACH INDUSTRIES 1932-2000	ISBN 1-58388-039-9
FLXIBLE TRANSIT BUSES 1953-1995	ISBN 1-58388-053-4
THE GENERAL MOTORS NEW LOOK BUS	ISBN 1-58388-007-0
GREYHOUND BUSES 1914-2000	ISBN 1-58388-027-5
MACK® BUSES 1900-1960 *	ISBN 1-58388-020-8
TRAILWAYS BUSES 1936-2001	ISBN 1-58388-029-1
TROLLEY BUSES 1913-2001	ISBN 1-58388-057-7
YELLOW COACH BUSES 1923-1943	ISBN 1-58388-054-2

EMERGENCY VEHICLES
AMERICAN LAFRANCE 700 SERIES 1945-1952	ISBN 1-882256-90-5
AMERICAN LAFRANCE 700 SERIES 1945-1952 VOLUME 2	ISBN 1-58388-025-9
AMERICAN LAFRANCE 700 & 800 SERIES 1953-1958	ISBN 1-882256-91-3
AMERICAN LAFRANCE 900 SERIES 1958-1964	ISBN 1-58388-002-X
CROWN FIRECOACH 1951-1985	ISBN 1-58388-047-X
CLASSIC AMERICAN AMBULANCES 1900-1979	ISBN 1-882256-94-8
CLASSIC AMERICAN FUNERAL VEHICLES 1900-1980	ISBN 1-58388-016-X
CLASSIC SEAGRAVE 1935-1951	ISBN 1-58388-034-8
FIRE CHIEF CARS 1900-1997	ISBN 1-882256-87-5
HEAVY RESCUE TRUCKS 1931-2000	ISBN 1-58388-045-3
INDUSTRIAL AND PRIVATE FIRE APPARATUS 1925-2001	ISBN 1-58388-049-6
LOS ANGELES CITY FIRE APPARATUS 1953 - 1999	ISBN 1-58388-012-7
MACK MODEL C FIRE TRUCKS 1957-1967 *	ISBN 1-58388-014-3
MACK MODEL CF FIRE TRUCKS 1967-1981 *	ISBN 1-882256-63-8
MACK MODEL L FIRE TRUCKS 1940-1954 *	ISBN 1-882256-86-7
MAXIM FIRE APPARATUS 1914-1989	ISBN 1-58388-050-X
NAVY & MARINE CORPS FIRE APPARATUS 1836 -2000	ISBN 1-58388-031-3
PIERCE ARROW FIRE APPARATUS 1979-1998	ISBN 1-58388-023-2
POLICE CARS: RESTORING, COLLECTING & SHOWING AMERICA'S FINEST SEDANS	ISBN 1-58388-046-1
SEAGRAVE 70TH ANNIVERSARY SERIES	ISBN 1-58388-001-1
VOLUNTEER & RURAL FIRE APPARATUS	ISBN 1-58388-005-4
WARD LAFRANCE FIRE TRUCKS 1918-1978	ISBN 1-58388-013-5
WILDLAND FIRE APPARATUS 1940-2001	ISBN 1-58388-056-9
YOUNG FIRE EQUIPMENT 1932-1991	ISBN 1-58388-015-1

RACING
EL MIRAGE IMPRESSIONS: DRY LAKES LAND SPEED RACING	ISBN 1-58388-059-3
GT40	ISBN 1-882256-64-6
INDY CARS OF THE 1950s, LUDVIGSEN LIBRARY SERIES	ISBN 1-58388-018-6
INDY CARS OF THE 1960s, LUDVIGSEN LIBRARY SERIES	ISBN 1-58388-052-6
INDIANAPOLIS RACING CARS OF FRANK KURTIS 1941-1963	ISBN 1-58388-026-7
JUAN MANUEL FANGIO WORLD CHAMPION DRIVER SERIES	ISBN 1-58388-008-9
LE MANS 1950: THE BRIGGS CUNNINGHAM CAMPAIGN	ISBN 1-882256-21-2
MARIO ANDRETTI WORLD CHAMPION DRIVER SERIES	ISBN 1-58388-009-7
NOVI V-8 INDY CARS 1941-1965 LUDVIGSEN LIBRARY SERIES	ISBN 1-58388-037-2
SEBRING 12-HOUR RACE 1970	ISBN 1-882256-20-4
VANDERBILT CUP RACE 1936 & 1937	ISBN 1-882256-66-2

RAILWAYS
CHICAGO, ST. PAUL, MINNEAPOLIS & OMAHA RAILWAY 1880-1940	ISBN 1-882256-67-0
CHICAGO & NORTH WESTERN RAILWAY 1975-1995	ISBN 1-882256-76-X
GREAT NORTHERN RAILWAY 1945-1970	ISBN 1-882256-56-5
GREAT NORTHERN RAILWAY 1945-1970 VOL 2	ISBN 1-882256-79-4
MILWAUKEE ROAD 1850-1960	ISBN 1-882256-61-1
MILWAUKEE ROAD DEPOTS 1856-1954	ISBN 1-58388-040-2
SHOW TRAINS OF THE 20TH CENTURY	ISBN 1-58388-030-5
SOO LINE 1975-1992	ISBN 1-882256-68-9
TRAINS OF THE TWIN PORTS, DULUTH-SUPERIOR IN THE 1950s	ISBN 1-58388-003-8
TRAINS OF THE CIRCUS 1872-1956	ISBN 1-58388-024-0
TRAINS of the UPPER MIDWEST STEAM&DIESEL in the1950S&1960S	ISBN 1-58388-036-4
WISCONSIN CENTRAL LIMITED 1987-1996	ISBN 1-882256-75-1
WISCONSIN CENTRAL RAILWAY 1871-1909	ISBN 1-882256-78-6

TRUCKS
BEVERAGE TRUCKS 1910-1975	ISBN 1-882256-60-3
BROCKWAY TRUCKS 1948-1961 *	ISBN 1-882256-55-7
CHEVROLET EL CAMINO INCL GMC SPRINT & CABALLERO	ISBN 1-58388-044-5
CIRCUS AND CARNIVAL TRUCKS 1923-2000	ISBN 1-58388-048-8
DODGE PICKUPS 1939-1978	ISBN 1-882256-82-4
DODGE POWER WAGONS 1940-1980	ISBN 1-882256-89-1
DODGE POWER WAGON	ISBN 1-58388-019-4
DODGE RAM TRUCKS 1994-2001	ISBN 1-58388-051-8
DODGE TRUCKS 1929-1947	ISBN 1-882256-36-0
DODGE TRUCKS 1948-1960	ISBN 1-882256-37-9
FORD HEAVY-DUTY TRUCKS 1948-1998	ISBN 1-58388-043-7
JEEP 1941-2000	ISBN 1-58388-021-6
JEEP PROTOTYPES & CONCEPT VEHICLES	ISBN 1-58388-033-X
LOGGING TRUCKS 1915-1970	ISBN 1-882256-59-X
MACK MODEL AB *	ISBN 1-882256-18-2
MACK AP SUPER-DUTY TRUCKS 1926-1938 *	ISBN 1-882256-54-9
MACK MODEL B 1953-1966 VOL 1 *	ISBN 1-882256-19-0
MACK MODEL B 1953-1966 VOL 2 *	ISBN 1-882256-34-4
MACK EB-EC-ED-EE-EF-EG-DE 1936-1951 *	ISBN 1-882256-29-8
MACK EH-EJ-EM-EQ-ER-ES 1936-1950 *	ISBN 1-882256-39-5
MACK FC-FCSW-NW 1936-1947 *	ISBN 1-882256-28-X
MACK FG-FH-FJ-FK-FN-FP-FT-FW 1937-1950 *	ISBN 1-882256-35-2
MACK LF-LH-LJ-LM-LT 1940-1956 *	ISBN 1-882256-38-7
MACK TRUCKS *	ISBN 1-882256-88-3
NEW CAR CARRIERS 1910-1998	ISBN 1-882256-98-0
PLYMOUTH COMMERCIAL VEHICLES	ISBN 1-58388-004-6
REFUSE TRUCKS	ISBN 1-58388-042-9
STUDEBAKER TRUCKS 1927-1940	ISBN 1-882256-40-9
STUDEBAKER TRUCKS 1941-1964	ISBN 1-882256-41-7
WHITE TRUCKS 1900-1937	ISBN 1-882256-80-8

TRACTORS & CONSTRUCTION EQUIPMENT
CASE TRACTORS 1912-1959	ISBN 1-882256-32-8
CATERPILLAR	ISBN 1-882256-70-0
CATERPILLAR POCKET GUIDE THE TRACK-TYPE TRACTORS 1925-1957	ISBN 1-58388-022-4
CATERPILLAR D-2 & R-2	ISBN 1-882256-99-9
CATERPILLAR D-8 1933-1974 INCLUDING DIESEL 75 & RD-8	ISBN 1-882256-96-4
CATERPILLAR MILITARY TRACTORS VOLUME 1	ISBN 1-882256-16-6
CATERPILLAR MILITARY TRACTORS VOLUME 2	ISBN 1-882256-17-4
CATERPILLAR SIXTY	ISBN 1-882256-05-0
CATERPILLAR TEN INCLUDING 7C FIFTEEN & HIGH FIFTEEN	ISBN 1-58388-011-9
CATERPILLAR THIRTY 2ND ED. INC. BEST THIRTY, 6G THIRTY & R-4	ISBN 1-58388-006-2
CLETRAC AND OLIVER CRAWLERS	ISBN 1-882256-43-3
CLASSIC AMERICAN STEAMROLLERS 1871-1935	ISBN 1-58388-038-0
FARMALL CUB	ISBN 1-882256-71-9
FARMALL F- SERIES	ISBN 1-882256-02-6
FARMALL MODEL H	ISBN 1-882256-03-4
FARMALL MODEL M	ISBN 1-882256-15-8
FARMALL REGULAR	ISBN 1-882256-14-X
FARMALL SUPER SERIES	ISBN 1-882256-49-2
FORDSON 1917-1928	ISBN 1-882256-33-6
HART-PARR	ISBN 1-882256-08-5
HOLT TRACTORS	ISBN 1-882256-10-7
INTERNATIONAL TRACTRACTOR	ISBN 1-882256-48-4
INTERNATIONAL TD CRAWLERS 1933-1962	ISBN 1-882256-72-7
JOHN DEERE MODEL A	ISBN 1-882256-12-3
JOHN DEERE MODEL B	ISBN 1-882256-01-8
JOHN DEERE MODEL D	ISBN 1-882256-00-X
JOHN DEERE 30 SERIES	ISBN 1-882256-13-1
MINNEAPOLIS-MOLINE U-SERIES	ISBN 1-882256-07-7
OLIVER TRACTORS	ISBN 1-882256-09-3
RUSSELL GRADERS	ISBN 1-882256-11-5
TWIN CITY TRACTOR	ISBN 1-882256-06-9

*This product is sold under license from Mack Trucks, Inc. Mack is a registered Trademark of Mack Trucks, Inc. All rights reserved.

All Iconografix books are available from direct mail specialty book dealers and bookstores worldwide, or can be ordered from the publisher. For book trade and distribution information or to add your name to our mailing list and receive a **FREE CATALOG** contact:

Iconografix, PO Box 446, Hudson, Wisconsin, 54016
Telephone: (715) 381-9755, (800) 289-3504 (USA),
Fax: (715) 381-9756

MORE GREAT TITLES FROM ICONOGRAFIX

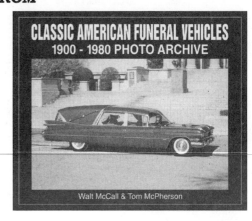

FIRE CHIEF CARS 1900-1997 PHOTO ARCHIVE
ISBN 1-882256-87-5

CLASSIC AMERICAN FUNERAL VEHICLES 1900-1980 PHOTO ARCHIVE
ISBN 1-58388-016-X

CLASSIC AMERICAN AMBULANCES 1900-1979 PHOTO ARCHIVE
ISBN 1-882256-94-8

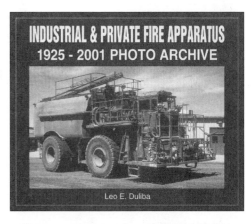

INDUSTRIAL & PRIVATE FIRE APPARATUS 1925-2001 PHOTO ARCHIVE
ISBN 1-58388-049-6

CROWN FIRECOACH 1951-1985 PHOTO ARCHIVE
ISBN 1-58388-047-X

MAXIM FIRE APPARATUS 1914-1989 PHOTO ARCHIVE
ISBN 1-58388-050-X

VOLUNTEER & RURAL FIRE APPARATUS PHOTO GALLERY
ISBN 1-58388-005-4

NAVY & MARINE CORPS FIRE APPARATUS 1836-2000 PHOTO GALLERY
ISBN 1-58388-031-3

POLICE CARS: RESTORING, COLLECTING & SHOWING AMERICA'S FINEST SEDANS
ISBN 1-58388-046-1

ICONOGRAFIX,
PO BOX 446, HUDSON, WISCONSIN, 54016
FOR A FREE CATALOG CALL:
1-800-289-3504

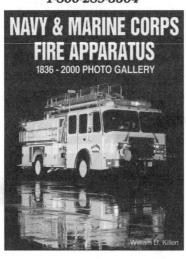

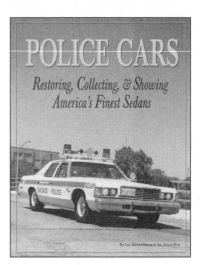